SELECTED READINGS IN PHYSICS

General Editor: D. TER HAAR

MEN OF PHYSICS

BENJAMIN THOMPSON – COUNT RUMFORD

Count Rumford on the Nature of Heat

COUNT RUMFORD

(*by permission of the Royal Institution, London*)

Cover Illustration by kind permission of Harvard University.
Bequest of Edmund Cogswell Converse.

MEN OF PHYSICS

BENJAMIN THOMPSON—COUNT RUMFORD

Count Rumford on the Nature of Heat

BY

SANBORN C. BROWN

Massachusetts Institute of Technology

PERGAMON PRESS

OXFORD · LONDON · EDINBURGH · NEW YORK

TORONTO · SYDNEY · PARIS · BRAUNSCHWEIG

PERGAMON PRESS LTD.,
Headington Hill Hall, Oxford
4 & 5 Fitzroy Square, London W.1

PERGAMON PRESS (SCOTLAND) LTD.,
2 & 3 Teviot Place, Edinburgh 1

PERGAMON PRESS INC.,
44–01 21st Street, Long Island City, New York 11101

PERGAMON OF CANADA LTD.,
6 Adelaide Street East, Toronto, Ontario

PERGAMON PRESS (AUST.) PTY. LTD.,
20–22 Margaret Street, Sydney, New South Wales

PERGAMON PRESS S.A.R.L.,
24 rue des Écoles, Paris 5e

VIEWEG & SOHN GmbH,
Burgplatz 1, Braunschweig

First edition 1967
Library of Congress Catalog Card No. 66–28414

Printed in Great Britain by A. Wheaton & Co. Ltd., Exeter

(3108/67)

CONTENTS

PREFACE

THE publication explosion in scientific literature had dictated such an economy of publication space that the periodical literature in physics at the present time is written in a peculiar type of clipped language and laconic style which conveys nothing but the bare outline of the scientific contribution. Any attempt to produce an interesting story or to speculate on the value or implications of the contribution would be ruthlessly edited from a manuscript if it were to be accepted for publication in any of the scientific journals at the present time.

This has not always been the case and in the eighteenth and early nineteenth centuries the accepted scientific writings were often attempts at literary efforts which were aimed at being as interesting to the general public as it was useful to the professional natural philosopher. At the present time, the only professionally accepted equivalent to eighteenth-century scientific reporting is the colloquium talk or the occasional hour-long invited paper to professional society. In this process of streamlining and making efficient scientific reporting, something of very real value has been lost which shows up in many ways, including the currently popular image of the scientific endeavor as one of inhuman attention to "cold fact" and the reputation of the scientific enterprise as one in which the scientist himself, as a human being, must never appear. The fact is, of course, that the scientist today is just as human as the scientist of 150 to 200 years ago but until we find some better way of transmitting scientific information than the present written scientific communication to cope with the information explosion of the modern scientific age, the current tendency will continue toward making the scientific paper less and less readable to more and more people.

It is very difficult for the nonscientist of the present day to develop an adequate picture of the scientific enterprise. As we have just mentioned, the scientific literature of today uses a kind of professional shorthand which has communication significance only to the trained professional. Furthermore, the prerequisite background education necessary to understand and evaluate scientific advancements which are of current interest to the research world in physics is of such a detailed and specific nature as to make modern examples of scientific methodology difficult, if not impossible, to capitalize upon. On the other hand, good scientific work is by no means limited to the modern idiom and the better examples of the operation of scientists in trying to understand the laws of nature can more often be found in the historical perspective than in today's laboratories and literature. With this in mind, the current volume has been assembled.

Count Rumford typifies much of the scientific enterprise of the eighteenth century. Few of the leading research scientists were trained professionals. The teaching profession was a rigidly disciplined and, in general, poorly paid profession which allowed the teachers neither the time nor the energy for independent investigation. The necessary background to arrive at the leading edge of science could be assimilated by anyone who had the time and inclination to study the then rather meager literature on the subject of physics so that the whole social structure of the scientific enterprise encouraged progress in natural philosophy being made by those not primarily engaged in educating the youth. It was an age in which the learned society fulfilled the function of our present-day research laboratory and the scientific communication was written to be interesting to the general educated world. Freed from any restriction on length or the efficiency of information transmission, the eighteenth-century physicist could describe not only his scientific contribution but the manner in which he was led to the subject matter under discussion. Thus, a much clearer picture emerged of the actual progress of scientific research than is now to be gained by reading the professional literature dominated as it is by the requirement for efficiency and sterile pedantry.

You should not only read these scientific discourses as interesting reports of scientific development but hopefully the character of the scientific enterprise and the scientists involved will be more realistically portrayed. If you are interested in the extraordinary life of Count Rumford himself, a short biography entitled "Count Rumford" by Sanborn C. Brown was published by Doubleday and Co. Inc., Garden City, New York, in 1962 and by Heinemann Educational Books, Ltd. of London in 1964.

PART 1

A BIOGRAPHICAL SKETCH OF COUNT RUMFORD, BENJAMIN THOMPSON (1753–1814)

THE details of a man's life can really never be separated from his contributions to society be they the work of his hands or the work of his brain. However, it is not practical in a volume of this sort to combine a reprinting of Rumford's scientific works with the details of his life which obviously interacted with his scientific endeavors. Nevertheless it seems worthwhile to sketch out the kind of life he led and the type of personality the author of these scientific papers exhibited to the world around him.

Count Rumford was born with the name Benjamin Thompson on March 23, 1753, in Woburn, Massachusetts, a small village close to Boston. There was nothing particularly noteworthy about his childhood, but he came to the attention of the public in his early twenties by actively working for the British Crown at a time when those who were actually rebelling against the British rule were successfully making it not only unhealthy but physically dangerous for Loyalists to stay in the New England area. There were, of course, many people in the American Colonies that were loyal to George III, but the details of their lives were usually obscured by the simple fact that they fled to either Nova Scotia or England and their property was confiscated by the victorious Americans, or they were farmers, shopkeepers and artisans, the records of whose lives often go unrecorded. This is not the case, however, with Benjamin Thompson through a series of rather remarkable events.

Throughout his life Benjamin Thompson demonstrated a consistent ability to capitalize upon available circumstances to further his own ends. He first catapulted himself into the public eye when at the age of 19 he married a wealthy widow, fourteen years his senior, and thereby became one of the wealthiest landholders in what is now the State of New Hampshire. Sarah Rolfe had married a man much older than herself who was a colonel in the Royal Governor's Militia and an honored member of Governor Wentworth's entourage. It was, therefore, not surprising that when young Benjamin Thompson took Colonel Rolfe's widow as a wife he should find easy access to the fashionable society which gravitated about the Royal Governor's establishments.

As a mark of Governor Wentworth's approbation of the marriage, not only did the Governor sign the marriage contract himself, but made the young lad a major in the Colonial Militia. Major Thompson, for his part, returned excellent service to the Loyalist cause by organizing a technique for rounding up deserters from the Regular British Army with such efficiency that his zeal was commended in dispatches from the Colonies to the Earl of Dartmouth in London.

As the course of revolution built up toward its inevitable climax of open hostility, Thompson's position as an active Loyalist increased his physical danger to an extent that in the winter of 1774, after having been openly accused of activities inimical to the cause of "American liberty", he fled from Concord, New Hampshire, and returned to Woburn, Massachusetts, and his family home, leaving his wife and an infant daughter to pacify an angry citizenry.

While Major Thompson was in Woburn ostensibly to get his affairs in order so that he could become an officer in George Washington's army, actually he continued his service to the Crown as a spy on the American establishment. Here he demonstrated two traits which were characteristic of his entire life. One of these was the fact that he was a very careful observer and recorded what he saw with great precision and attention to detail. This characteristic is amply borne out in his scientific papers and will be evident to all who read this volume. Using his Woburn home as a base, Benjamin Thompson wrote a most careful and detailed document for the British Military Establishment entitled "Observations on the Present State of the Rebel Army", which not only chronicled the physical and military details of the forces under Washington's command in Cambridge, but discussed problems of morale and tactics in a most perceptive and useful (to the British) way.

The other facet which stands out so clearly in his activities at this time in Woburn was the very practical application of his scientific learning. Thompson was the author of the first known intelligence written in secret ink in the American Revolution, and his technical excellence in this area was not exceeded until the First World War. It was obvious that he had so much faith in its superiority that he was willing to risk his life on the security of his secret ink letters, and his whole approach to this use of the then most modern technology was a tribute to his ability as an experimental scientist.

Although Thompson's science and technology was excellent enough so that his espionage activities were never discovered by his contemporaries, his behavior in the eyes of the citizens of Woburn and Cambridge was not above suspicion, and after being examined twice by local "Committees of Safety", Benjamin Thompson decided that he could no longer trust his luck in the American camp and he fled to the safety of the British Army in Boston in October 1775.

When the military position of the British Army in Boston became untenable, the major part of the forces embarked for the safety of the Loyalist city of Halifax, Nova Scotia, but Thompson, as one of the most knowledgeable observers of the American situation, was attached to that small group of experts who went directly to London to report on the military situation to Lord George Germain, whose responsibility it was to crush the rebellious Colonies for George III.

London was in desperate need of reliable information about what was going on in the Massachusetts Bay Colony, and when the aggressively intelligent and well-informed Thompson appeared on the scene, he was such a welcome addition to the government bureaucracy that he rose rapidly in the favor of the King's ministers.

Benjamin Thompson landed in London during the summer of 1776 and almost immediately took up a position as private secretary to Lord George Germain, Secretary of State for the Colonies. Within three years Thompson had made himself so useful to the British Government that he received the position of Secretary of the Province of Georgia, and in 1780 he was made Under Secretary of State for the Northern Department. As he rose in government circles, he found time not only to be a useful political figure, but also to gain a reputation as a serious natural philosopher which earned him, in 1781, Fellowship in the Royal Society. The research work which won him this honor was a series of studies of the force of fired gunpowder which he carried out in the summer of 1778 on the summer estate of Lord George Germain. Thompson perfected a method of testing the force of gunpowder by means of a ballistic pendulum which is still a common physics demonstration in using the conservation of momentum for measuring the velocity of rifle bullets.

Thompson was professionally considered a military man and his ballistics studies greatly increased his reputation. However, it was not just the engineering aspect of gunnery that interested him. Even at the age of 25, he was searching for a clue as to the nature of heat, and many of his speculations on the nature of the force behind an explosion of gunpowder centered around a search for the explanation of the nature of heat itself.

The summer following his experimental studies at Lord George Germain's Stoneland Lodge he went on a three-months' tour of duty with Admiral Hardy's fleet in the English Channel, and here not only did he send back to Lord George detailed accounts of the inefficiency of the operation of the British fleet, but he carried out an extensive continuation of his ballistics studies with the great guns of the fleet at his disposal.

So well was he doing in his rise to fame and fortune that it is amazing to discover that in 1781 he suddenly left his position in London and sailed to take up an active command of the King's American Dragoons in the area of New York. Local gossip at the time linked his name with a notorious French spy who had been apprehended with naval secrets on his person, but no proof of the contemporary gossip has been found in any official documents. Whatever the reason, it is certainly true that he gave up a promising career for rather doubtful service in the Colonies.

Major Thompson's interlude with the King's American Dragoons just at the close of the American war was neither an important nor a particularly productive era of his life. It did, however, serve to consolidate his claim as a professional soldier, and he was able to use his experience not only to advance to a rank of colonel, but to retire not as a Colonial, but as a member of the Regular British Military Establishment. In this position, as he went to the Continent as a soldier of fortune late in 1783, he was sure of a position of honor, and he was not in any particular hurry to find a permanent position. In the spring of 1784 he investigated the possibility of becoming an aide-de-camp to the Elector of Bavaria, and after a brief return to London to be knighted by George III, so that his position in the English hierarchy would not be lower than his proposed standing in the Bavarian Court, he settled down to a long and productive career in the Bavarian military organization.

It was in Bavaria that Benjamin Thompson carried out the major part of his scientific work on the nature of heat. His position at the Bavarian Court was sufficiently high to assure not only the financial backing necessary for his philosophic investigations, but he also had at his disposal artisans, mechanics, and technical assistants who provided the necessary back-up for his varied and at times, fairly grandiose experimental investigations.

The Elector Karl Theodor was extremely impressed by his brilliant protégé, promoted him rapidly to positions of great power and responsibility, and in return Thompson did much to enhance the fame and fortune of the Bavarian Court not only in terms of his scientific and technical advances, but also in terms of his organizational innovations in the army, the educational system, and the social structure of the Bavarian economy. In grateful recognition to all he did for the Court at Munich, Thompson was honored in many ways not the least of which was to be made a

Count of the Holy Roman Empire in 1793, and in recognition of his start toward fame and fortune in Concord, New Hampshire, he took the ancient name of that New England village, Rumford, as his title. However, so many innovations were introduced by the energetic Count that the established order rebelled at the introduction of so much novelty and the voice of his enemies eventually outweighed those of his friends, and the Bavarian Elector was forced to find a way to relieve his protégé of his important place in the Bavarian Court.

The honorific situation which Elector Karl Theodor arranged for Count Rumford was that of Minister Plenipotentiary to the English Court. Rumford was duly pleased and packing bag and baggage, left Munich on the long and arduous trek to London. A serious flaw in the arrangements, however, was painfully evident when the Count arrived in London. The Elector had failed to ask permission of George III, who adamantly refused to accept the Count as a Bavarian Minister, and Rumford found himself in London without a job.

Perhaps for the course of science this was a fortunate affair, although for the Count it was a bitter blow. For what Rumford busied himself with after he recovered from his disappointment was to set up the now world-famous Royal Institution in London, not in its present form, but as a sort of museum of science and school for mechanics. It soon became evident that this original idea was not one which attracted the necessary financial support, and gradually through the years the museum and technical school aspect gave way to its present emphasis on popular science lectures and research. There is no doubt, however, that Rumford's organizational ability and vision in creating the Royal Institution stands as one of the enduring memorials to his energy, and his employment at the Royal Institution of such natural philosophers as Thomas Young and Humphrey Davy gives us the feeling that he was able to recognize real genius in his employees.

Count Rumford was an extremely egotistical and domineering figure, and it was not long before the Managers of the Royal Institution became thoroughly annoyed with his high-handed and at times arbitrary handling of the Institution's affairs. After numerous battles with the Managers, Count Rumford finally became so angry that he not only left the Royal Institution, but left England altogether and settled down in one country which was in a state of war with England, believing that Napoleon would give him a more friendly atmosphere in which to pursue his philosophic researches than he was able to create either in England or Bavaria.

For a time this appeared to be the case. He was lionized by French society, honored by the French Academies, and even married a famous French lady, the widow of the celebrated chemist Lavoisier. These idyllic surroundings, however, gradually faded as he fought scientifically with Laplace and Lagrange and discovered Madame Lavoisier so incompatible that they ultimately separated. Rumford died and is buried in what was then a suburb of Paris, the village of Auteuil. His death actually took place in August 1814.

INTRODUCTION

To ENGAGE in experiments on heat was always one of my most agreeable employments. This subject had already begun to excite my attention, when, in my seventeenth year, I read Boerhaave's admirable Treatise on Fire. Subsequently, indeed, I was often prevented by other matters from devoting my attention to it, but whenever I could snatch a moment I returned to it anew, and always with increased interest. Even now this object of my speculations is so present to my mind, however busy I may be with other affairs, that everything taking place before my eyes, having the slightest bearing upon it, immediately excites my curiosity and attracts my attention.

This habit of many years' standing, by force of which I seize with the greatest eagerness, and endeavour to investigate, each and every phenomenon related even in the slightest manner to heat and its operations which comes to my knowledge, has suggested to me almost all the experiments that I have performed with reference to this subject.

Count Rumford 1804
(Memoires sur la Chaleur)

CHAPTER I

PERSPECTIVE

THE name of Count Rumford is found in almost every textbook on general physics. Generally, however, one finds mention only of his cannon-boring experiments and a paraphrase of Professor John Tyndall's comment:† "Rumford, in this memoir annihilates the material theory of heat. Nothing more powerful on the subject has since been written."

The serious student is puzzled by the statement that a single experiment could bring about the complete downfall of the well-established caloric theory, or that one set of results should credit Count Rumford with having led the first successful attack on the accepted theory of heat of his day. The answer is, of course, that Rumford's great contemporary reputation was by no means based solely on this experiment with heat from friction, but on a long series of experiments extending over a period of thirty years, which attacked the caloric theory from many different points of view.

Rumford originated neither the energy theory of heat nor some of the experiments which he carried out to prove this theory. The work of Boyle, Hooke, Leibnitz and Locke seems all to have had its influence on Rumford's achievement. Rumford's most famous experiment is similar in principle to one reported by Boyle, and he practically quotes Hooke in some of his descriptions of heat as a form of energy. Nevertheless, his outstanding experimental skill, his endless search for ways to attack the caloric theory, as well as his international fame as a soldier, statesman and philosopher,

† Tyndall, John, *Heat Considered as a Mode of Motion*, p. 39, Appleton, N.Y., 1871.

made him one of the first successful advocates of our modern concept of heat.

At the end of the eighteenth century the two opposing theories of the nature of heat were being vigorously debated. The caloric theory explained the nature of heat as a fluid, and the energy theory considered it "a mode of motion". To appreciate Count Rumford's preeminence as an advocate of the energy theory, one must study the contemporary caloric theory in detail and consider what experimental tests led Rumford to disbelieve the existence of heat as a material fluid. Also we must understand what Rumford himself was puzzled about, and what he thought he was trying to prove. He wrote:†

> I must confess that it has always been impossible for me to explain the results of such experiments except by taking refuge in the very old doctrine which rests on the supposition that heat is nothing but a vibratory motion taking place among the particles of the body.

It has become common, in histories of science, to credit Count Rumford with founding the modern concept of heat. This is not the case. In the passage quoted just above, we find him "taking refuge in the very old" theory of heat as a form of energy. He did not even consider that he performed particularly original experiments in his generation of heat by friction, for on November 8, 1797 he wrote from Munich to Professor Pictet in Geneva:‡

> The results of my Experiments seem to me to prove to a demonstration that there is no such thing as an *Igneous fluid*, and consequently that Caloric has no real existence. You must not however raise your expectations too high respecting my experiments. Though they were made on a large scale, and conducted with care, there was nothing very new or very remarkable about them; and as to their results, they proved only this single fact—that the heat generated by friction is *inexhaustible*, even when the bodies rubbed together are, to all appearances, perfectly insulated, or put into a situation in which it is evidently impossible for them to receive from other bodies the heat they are continually giving off.
>
> It appears to me that that which any insulated body, or system of bodies can continue to give off without limitation, cannot be a *material substance*.

† Rumford, Count, *Memoires sur la Chaleur*, Paris, 1804.

‡ The American Academy of Arts and Sciences (Boston) has a collection of copies of letters from Rumford to Pictet.

Count Rumford made the first measurement which could be interpreted as the mechanical equivalent of heat. In connection with this subject one immediately thinks of James Prescott Joule who wrote:†

> One of the most important parts of Count Rumford's paper, though one to which little attention has hitherto been paid, is that in which he makes an estimate of the quantity of mechanical force required to produce a certain amount of heat. . . . According to Count Rumford's experiment . . . the heat required to raise 1 lb of water 1° will be equivalent to the force represented by 1034 foot-pounds. This result is not very widely different from that which I have deduced from my own experiments related in this paper, *viz.* 772 foot-pounds; and it must be observed that the excess of Count Rumford's equivalent is just such as might have been anticipated from the circumstance, which he himself mentions, that "no estimate was made of the heat accumulated in the wooden box, nor of that dispersed during the experiment."

Joule's contemporary as well as subsequent reputation was tremendous, particularly in the area of the mechanical equivalent of heat. This statement which has just been cited from Joule's writings has often been misinterpreted to prove that Rumford himself had the idea of a mechanical equivalent. This is not true, as a reading of his works will show. It was Joule who took Rumford's data to calculate this quantity. Rumford never refers to this concept himself.

Rumford's own theory of heat was a vibratory one, as will be shown later in this volume, and this hypothesis did not require a conservation of energy point of view, so that the whole issue of an energy equivalence was never raised.

† Joule, J. P., "On the mechanical equivalent of heat", *Phil. Trans.*, 1850, pp. 61–82.

CHAPTER 2

THE CALORIC THEORY

To APPRECIATE Count Rumford's preeminence as an advocate of an "anti-caloric" theory of heat, one must study the contemporary caloric theory in detail and consider what experimental tests led Rumford to disbelieve the existence of heat as a material fluid.

The fundamental assumption of the caloric theory was the concept that heat consisted of a fluid capable of penetrating all space and able to flow in and out of all substances. This fluid was called caloric. The important principles of its action were that it was self-repulsive and was strongly attracted by matter.

Matter was considered to be made up of atoms consisting of discrete particles attracted toward each other by their mutual gravitational attraction. If the gravitational attraction were the only force which occurred, every particle of matter would be attracted toward each other, resulting in a single solid homogeneous mass. To prevent this conclusion, a repulsive force was postulated which was considered to be the self-repulsive caloric.

Thermal Expansion

The caloric theory supplied an obvious solution to the problem of thermal expansion and contraction. Heating a body consisted of adding the fluid caloric to the body and it expanded. On cooling, the caloric fluid was removed and hence the body contracted. The detailed behavior of many of the phenomena of heat was explained by considering† each atom to be surrounded by an atmosphere of

† Emmett, *Annals of Philosophy* **9**, 421 (1817).

caloric whose density diminished more rapidly than the intensity of the gravitational attraction with distance from the center of any particle.

The gravitational attraction was considered to be inversely proportional to the square of the distance from the center of the atom while the caloric atmosphere which caused the repulsion was assumed to obey a logarithmic law by analogy to the earth's atmosphere. This might be illustrated as shown in Plate 1, where the gravitational attraction due to the atom *m* is represented as a

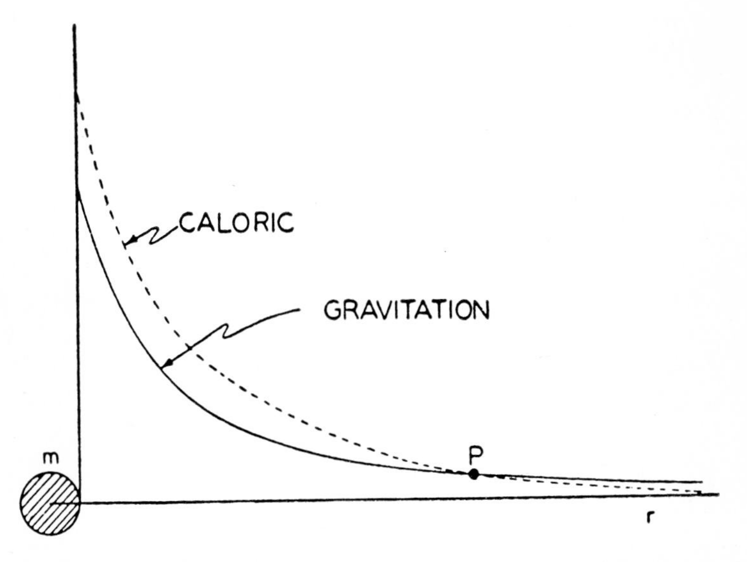

PLATE 1. Caloric repulsion and gravitational attraction between atoms. The gravitational attraction due to the atom *m* is represented by the solid line, and the caloric repulsion by the dotted one.

solid line and the caloric repulsion as dotted. At the point where they are equal, point *P*, another similar particle would be at equilibrium. If the temperature of the body of which *m* is an atom was increased, the caloric atmosphere around each atom increased and the caloric curve of Plate 1 was thought of as rising. When this occurred, point P receded from the center of m, corresponding to the expansion of the body. According to this picture, if the temperature of the body was increased in uniform steps, the height of the caloric atmosphere increased in a uniform fashion. This is illustrated in Plate 2. The theory predicted that the dilation of a

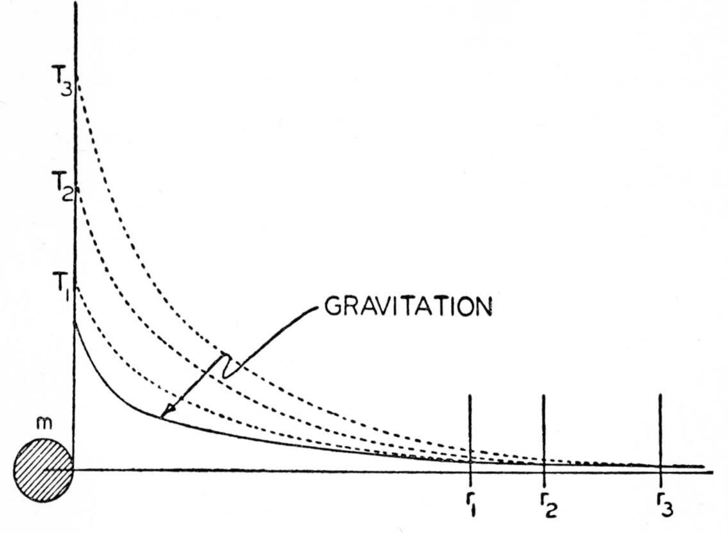

PLATE 2. Representation of the increase in dimensions of the caloric atmosphere with temperature.

body should not be a uniform function of the temperature but should increase with increasing temperature as was shown to be the case by the experiments of Dulong and Petit.†

The theoretical difference between solids, liquids, and gases was considered to lie in the degree of gravitational attraction between the atoms of the substance. With small amounts of heat, the caloric repulsion was not strong and the atoms were considered tightly bound by a strong gravitational attraction. As the temperature of the body was increased, the attraction became less as the repulsion

† Dulong and Petit, *Ann. de Chem. et de Physique* 7, 113 (1818).

became greater. In a liquid, the caloric content was sufficiently high so that the atoms were not held in a rigid position by the mutual gravitational attraction. In a gas the gravitational attraction was considered to be negligible. Thus the observed effect was predicted that the expansion of a gas would be much greater than in the case of a liquid, which in turn would be greater than that of a solid. The fact that the expansion coefficient increased with temperature much more rapidly for liquids than for solids, which was a result of this theory, had been demonstrated by the experiments of Lavoisier and Laplace.†

Proponents of this theory argued that the negligible effect of the gravitational attraction between atoms of a gas should be apparent in at least two ways. Since the gravitational effect depended upon the mass of the atoms, it was expected that solids and liquids differing in the mass of their atoms should have different expansion coefficients. Since for gases, the gravitational attraction was considered negligible, all gases should have the same thermal expansion coefficient. Also, the expansion of a gas per degree change in temperature should be independent of the actual temperature. Both of these effects had been found to be true for gases following the experiments of Gay-Lussac.‡

Specific Heat

A careful distinction was drawn between the intensity of heat and the quantity of heat. Plate 3 shows a graph representing the density of the caloric atmosphere around an atom, as a function of the distance from the atom. The two curves represent the caloric density at two different temperatures. The intensity of heat is represented by the temperature and therefore by the actual density of caloric at the surface of the atom. All atoms did not have identical calorific atmospheres, and although they all had a logarithmic dependence of caloric density on distance, the rate at which the atmospheric density falls off varies from substance to substance.

† Biot, *Traite de Physique*, vol. I, p. 158 (1816).
‡ Gay-Lussac, *Ann. de Chem.* **43,** 137 (1802).

The quantity of heat required to change a body from temperature T_1 to temperature T_2 is represented by the shaded difference between the two curves of Plate 3. Since these curves were considered

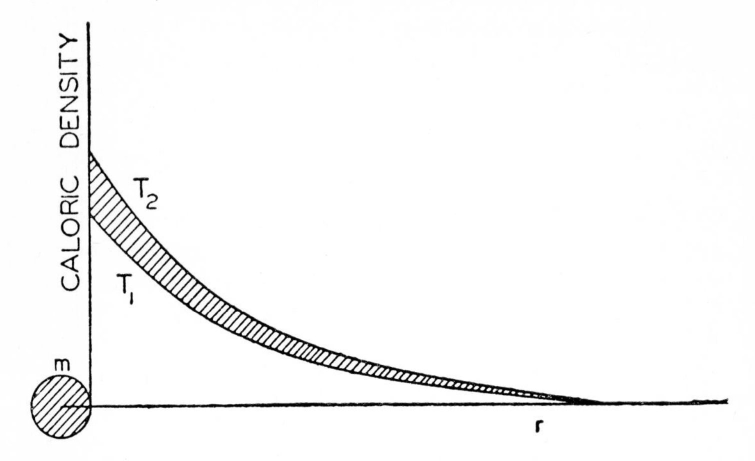

PLATE 3. The shaded area represents the quantity of heat required to change from temperature T_1 to the temperature T_2.

different for different types of atoms, the quantity of caloric involved in the same change in temperature for different types of material was expected to be different. This quantity of heat fluid required to produce a given change of temperature for a given amount of material was called the specific heat by analogy to the term specific gravity.†

The experiments of Dulong and Petit‡ showed that the product of the specific heat of each solid element by the weight of its atom gave a constant value. This led to the conclusion that the amount of caloric surrounding an atom was related to its atomic weight in a manner which was expected from the attraction between matter and caloric. In 1780 Laplace and Lavoisier showed that the specific heat of a substance was not a constant, but increased with

† Wilcke, *Proc. Roy. Soc. of Stockholm* (1772).
‡ Dulong and Petit, *Ann. de Chem. et de Physique* **10,** 395 (1819).

temperature. This agreed with the caloric theory. With reference to Plate 3, as the temperature increased in uniform increments, the difference in area between the successive caloric density curves increases, which predicted greater quantities of heat necessary for the change to take place, as was observed.

Changes of State

At a sufficiently low temperature, all known substances were solid. As the temperature was raised, caloric was attracted toward the atoms of the substance and eventually the substance liquefied. If more caloric was added, it would take the gaseous form. Black† showed that when a change of state occurred, the heat added to the body went only into the change in phase and the temperature of the body remained constant during the process. To explain this he introduced the concept of latent heat.

According to Black's theory, heat could take two different forms, sensible and latent. Changes in sensible heat corresponded to those changes which could be measured by changes in temperature. Besides this effect of the attraction between matter and caloric, in which the caloric merely forms an atmosphere around the atoms and molecules, caloric was considered actually to combine with an atom in a fashion similar to the chemical combinations of the atoms themselves. When this occurred, the caloric lost its sensible form and became latent. The "chemical" combination of an atom with caloric produced a new compound in which neither the original atom nor the caloric retained its identity. It took place, as did ordinary chemical reactions, only in definite proportions and under definite circumstances. No heat was considered lost in the process, since it was reversible; cooling a body down returned the caloric back to its sensible form.

No new explanation was necessary to describe the transition from liquid to gas. For the same substance, there seemed to be no particular relation between the latent heat of fusion and the latent

† Black, *Lectures on the Elements of Chemistry* (1803).

heat of vaporization, even though the same process was active in both. The case of vaporization, however, seemed to give an additional check on the validity of the caloric theory. The theory postulated that the requirement for a reaction to take place between the sensible caloric and an atom was a critical amount of caloric around each atom. Since caloric was self-repulsive, the temperature at which vaporization took place ought to be changed by changing the external pressure. When a piece of iron was compressed by hammering, sensible caloric was squeezed out and the surface of the iron became hot, or if gas was compressed it emitted caloric and became hotter. It seemed amply proved, therefore, that sensible caloric could be squeezed from a body by artificially pushing the atoms into a closer proximity than the mutual repulsion of their caloric atmospheres would allow. Thus, if pressure were put on a substance near its boiling point, some of the sensible caloric would be lost from the substance and a higher temperature would have to be applied before sufficient caloric was available to the atom for a vaporization reaction to occur. This fact had been amply borne out by experiment.

The Communication of Heat

Conduction was a very obvious method of transferring heat from one place to another because of the very great attraction between matter and caloric. The less caloric a body had, the greater the attraction of its atoms for the caloric fluid. In adding heat to one end of a solid bar, the atoms at the heated end acquired more caloric than their neighbors, and by having more, their attraction for this caloric was less. The neighboring atoms attracted the caloric away, and continued to do so until all the atoms of the substance had achieved the same caloric atmospheres. The facility with which caloric could be passed from one atom to another depended upon the structure and composition of the substance. In apparent agreement with the caloric theory, Count Rumford†

† Rumford, *Phil. Trans. Roy. Soc. (London)* **77,** 48 (1792).

found that even in the same substance, the conducting power increased with the compactness of the structure.

Count Rumford was the first to prove, in 1785, that heat could be transmitted through a vacuum. This is particularly interesting because the fact that heat could be transmitted through a vacuum without the aid of a material medium seemed to be one of the great stumbling blocks in the theory which described the nature of heat as a wave motion. It was a natural consequence of the caloric theory. Since caloric was a self-repulsive fluid which pervaded all space, it was to be expected that where there was no matter to impede its progress, it would expand without limit.

The radiation of heat was known to be strongly influenced by the nature and condition of the radiating surfaces. A perfectly polished surface was considered to have all the molecules lying in a plane, and all the molecules as close together as possible. Thus, since the greatest amount of mass was to be found at this kind of surface, the caloric was most tightly bound, and could be lost only with the greatest difficulty. This defined a poor radiator. That it was a good reflector would be predicted since there already existed a tightly bound caloric surface layer; this would repel any further caloric coming toward the surface. The good radiators were rough surfaces, which meant that for a given area of surface, much less mass lay in a plane and hence the caloric was held only by a few groups of atoms at any one place. The neighboring atoms did not have the cumulative effect present if there were more of them as neighbors, so the caloric could escape more easily and one obtained a good radiator.

Heat from Mechanical Work

When gas was put in a cylinder and compressed, heat was given off. The atoms of the gas were considered unaffected by the gravitational attraction of neighboring gas atoms, and held apart by the caloric atmospheres surrounding each atom. Pushing the atoms closer together could only be done by overcoming the thermorepulsion by mechanical force. In so doing, caloric was squeezed out of the gas and it gave off heat. If the reverse experiment was tried and the gas expanded, the temperature would fall since the atoms would be farther apart than their caloric atmospheres would keep them.

The removal of heat by compression was by no means confined to gaseous bodies. The action of compression on solid bodies afforded a further demonstration of the existence of the caloric fluid. It was noticed especially that caloric was disengaged by pressure or percussion only so long as bodies underwent condensation. A common experiment was to show that a piece of iron made red hot by hammering could not be strongly heated a second time unless it was reheated in the fire. This fact was explained by supposing that the fluid heat which had been pressed out of it by percussion was recovered in the fire.

Discussion

The caloric theory has a great influence on our teaching of the elementary concepts of heat. Much of the nomenclature in common use is based on the older theory: we talk about heat flowing from one place to another, and many text books discuss the concepts of specific heat and calorimetry essentially in terms of the caloric theory. The terms specific and latent heat had a descriptive meaning in the heat fluid theory, and the unit, and in fact the use of the words "quantity of heat", are caloric theory concepts. A careful study into the details of what is being taught at the elementary level concerning the behavior and phenomena of heat shows that to a large degree students are introduced to the subject by way of the caloric theory of heat and it is not until the more advanced courses that a real effort is made to present a consistent discussion of heat as a form of energy.

PART 2

CHAPTER 3

THE PROPAGATION OF HEAT IN FLUIDS

DURING his scientifically productive life, Count Rumford made many discoveries that profoundly affected the physical thought of his and succeeding generations. His research on the propagation of heat in fluids was aimed at understanding the nature of heat, and in particular, attempting to demonstrate flaws in the caloric theory.

The picture of heat conduction consistent with the fluid nature of caloric was an actual transport of the heat fluid from one place to another. Count Rumford's experiments in the transport of heat in a fluid were of three types. First he showed that heat could be transported from one place to another by the mass motion of the liquid itself. He then showed that if this motion did not take place the fluid appeared to be an insulator. Finally he demonstrated that even in a liquid at constant temperature there was an internal motion of the liquid particles which provided ever-ready carriers of heat in case a temperature gradient developed.

Convection

In the course of his first group of experiments, Rumford discovered and clearly demonstrated the existence of convection currents. He also studied, in a very full way, the effect on the transfer of heat through a liquid of interrupting the convective motion of the fluid material.

The propagation of heat by means of the internal motion of the fluid particles went without a name for many years. It was not until 1834 that William Prout, writing in one of the *Bridgewater Treatises*, suggested:†

> There is at present no single term in our language employed to denote this mode of propagation of heat; but we venture to propose for that purpose the term *convection* (*convectio*, a carrying or conveying), which not only expresses the leading facts, but also accords very well with the two other terms [conduction and radiation].

† *Bridgewater Treatises* (W. Pickering, London), vol. 8, p. 65. Volume 8, which was published in 1834, was written by Prout.

Even after this name was suggested, one finds that it was another 20 years before *convection* found its way to the universal acceptance which the term now enjoys.

ESSAY VII – PART 1

The Propagation of Heat in Fluids (Chapter 1)

Published separately by Cadell and Davies, London, 1797
Bibliotheque Britannique (Science et Arts) v, p. 90 and 97–200 (1797)
Nicholson's quarto Journal I p. 289–296, 341–348, 563–575 (Extracts) (1797)
Gilbert's Annalen der Physik I (1799) p. 214–241, 323–351, 436–463
Gren's Neues Journal der Physik IV p. 418–450 (1797)
Crell's Chemische Annalen, 1797, p. 78–104, 149–170, 233–246, 342–358, 446–464, 488–502

It is certain that there is nothing more dangerous in philosophical investigations, than to take anything for granted, however unquestionable it may appear, till it has been proved by direct and decisive experiment.

I have very often, in the course of my philosophical researches, had occasion to lament the consequences of my inattention to this most necessary precaution.

There is not, perhaps, any phenomenon that more frequently falls under our observation than the Propagation of Heat. The changes of the temperature of sensible bodies, of solids, liquids, and elastic fluids, are going on perpetually under our eyes, and there is no fact which one would not as soon think of calling in question as to doubt of the free passage of Heat, in all directions, through all kinds of bodies. But, however obviously this conclusion appears to flow from all that we observe and experience in the common course of life, yet it is certainly not true; and to the erroneous opinion respecting this matter, which has been universally entertained by the *learned* and by the *unlearned*, and which has, I believe, never even been called in question, may be attributed the little progress that has been made in the investigation of the science of Heat,—a science, assuredly of the utmost importance to mankind!

Under the influence of this opinion I, many years ago, began my experiments on Heat; and had not an accidental discovery drawn my attention with irresistible force and fixed it on the subject, I probably never should have entertained a doubt of the free passage of Heat *through air*; and even after I had found reason to conclude, from the results of experiments which to me appeared to be perfectly decisive, that air is a *non-conductor* of Heat, or that Heat cannot pass through it without being transported by its particles, which, in this process, act individually or independently of each other; yet, so far from pursuing the subject and contriving experiments to ascertain the manner in which Heat is communicated in other bodies, I was not sufficiently awakened to suspect it to be even possible that this quality could extend farther than to elastic Fluids.

With regard to liquids, so entirely persuaded was I that Heat could pass freely *in them* in all directions, that I was perfectly blinded by this prepossession, and rendered incapable of seeing the most striking and most evident proofs of the fallacy of this opinion.

I have already given an account, in one of my late publications (Essay on the Management of Fire and the Economy of Fuel), of the manner in which I was led to discover that *steam* and *flame* are *non-conductors* of Heat. I shall now lay before the public an account of a number of experiments I have lately made, which seem to show that *water*, and probably all other liquids, and Fluids of every kind, possess the same property. That is to say, that, although the particles of any Fluid, *individually*, can receive Heat from other bodies or communicate it to them, yet among these particles themselves all *interchange* and *communication* of Heat is absolutely impossible.

It may, perhaps, be thought not altogether uninteresting to be acquainted with the various steps by which I was led to an experimental investigation of this curious subject of enquiry.

When dining, I had often observed that some particular dishes retained their Heat much longer than others, and that apple-pies, and apples and almonds mixed (a dish in great repute in England), remained hot a surprising length of time.

Much struck with this extraordinary quality of retaining Heat which apples appeared to possess, it frequently occurred to my recollection; and I never burnt my mouth with them, or saw others meet with the same misfortune, without endeavouring, but in vain, to find out some way of accounting in a satisfactory manner for this surprising phænomenon.

About four years ago, a similar accident awakened my attention, and excited my curiosity still more: being engaged in an experiment which I could not leave, in a room heated by an iron stove, my dinner, which consisted of a bowl of thick rice-soup, was brought into the room, and as I happened to be too much engaged at the time to eat it, in order that it might not grow cold, I ordered it to be set down on the top of the stove; about an hour afterwards, as near as I can remember, beginning to grow hungry, and seeing my dinner standing on the stove, I went up to it and took a spoonful of the soup, which I found almost cold and quite thick. Going, by accident, deeper with the spoon the second time, this second spoonful burnt my mouth.* This accident recalled very forcibly to my mind the recollection of the hot apples and almonds with which I had so often burned my mouth a dozen years before in England; but even this, though it surprised me very much, was not sufficient to open my eyes, and to remove my prejudices respecting the conducting power of water.

Being at Naples in the beginning of the year 1794, among the many natural curiosities which attracted my attention, I was much struck with several very interesting phænomena which the hot baths of Baiæ presented to my observation, and among them there was one which quite astonished me: standing on the sea-shore near the baths, where the hot steam was issuing out of every crevice of the rocks, and even rising up out of the ground, I had

* It is probable that the stove happened to be nearly cold when the bowl was set down upon it, and that the soup had grown almost cold; when a fresh quantity of fuel being put into the stove, the Heat had been suddenly increased.

the curiosity to put my hand into the water. As the waves which came in from the sea followed each other without intermission, and broke over the even surface of the beach, I was not surprised to find the water cold; but I was more than surprised, when, on running the ends of my fingers through the cold water into the sand, I found the heat so intolerable that I was obliged instantly to remove my hand. The sand was perfectly wet, and yet the temperature was so very different at the small distance of two or three inches! I could not reconcile this with the supposed great conducting power of water. I even found that the top of the sand was, to all appearance, quite as cold as the water which flowed over it, and this increased my astonishment still more. I then, for the first time, began to doubt of the conducting power of water, and resolved to set about making experiments to ascertain the fact. I did not, however, put this resolution into execution till about a month ago, and should perhaps never have done it, had not another unexpected appearance again called my attention to it, and excited afresh all my curiosity.

In the course of a set of experiments on the communication of Heat, in which I had occasion to use thermometers of an uncommon size (their globular bulbs being above four inches in diameter) filled with various kinds of liquids, having exposed one of them, which was filled with spirits of wine, in as great a heat as it was capable of supporting, I placed it in a window, where the sun happened to be shining, to cool; when, casting my eye on its tube, which was quite naked (the divisions of its scale being marked in the glass with a diamond), I observed an appearance which surprised me, and at the same time interested me very much indeed. I saw the whole mass of the liquid in the tube in a most rapid motion, running swiftly in two opposite directions, *up* and *down* at the same time. The bulb of the thermometer, which is of copper, had been made two years before I found leisure to begin my experiments, and having been left unfilled, without being closed with a stopple, some fine particles of dust had found their way into it, and these particles, which were intimately mixed with the spirits of wine, on their being illuminated by the sun's beams, became perfectly visible (as the dust in the air of a darkened room is illuminated and rendered visible by the sunbeams which come in through a hole in the window-shutter), and by their motion discovered the violent motions by which the spirits of wine in the tube of the thermometer was agitated.

This tube, which is $\frac{43}{100}$ of an inch in diameter internally, and very thin, is composed of very transparent, colourless glass, which rendered the appearance clear and distinct and exceedingly beautiful. On examining the motion of the spirits of wine with a lens, I found that the ascending current occupied the *axis of the tube*, and that it descended by the *sides of the tube*.

On inclining the tube a little, the *rising* current moved out of the axis and occupied that side of the tube which was uppermost, while the *descending* current occupied the whole of the lower side of it.

When the cooling of the spirits of wine in the tube was hastened by wetting the tube with ice-cold water, the velocities of both the ascending and the descending currents were sensibly accelerated.

The velocity of these currents was gradually lessened as the thermometer was cooled, and when it had acquired nearly the temperature of the air of the room, the motion ceased entirely.

By wrapping up the bulb of the thermometer in furs, or any other warm covering, the motion might be greatly prolonged.

I repeated the experiment with a similar thermometer of equal dimensions, filled with linseed-oil, and the appearances, on setting it in the window to cool, were just the same. The directions of the currents, and the parts they occupied in the tube, were the same, and their motions were to all appearance quite as rapid as those in the thermometer which was filled with spirits of wine.

Having now no longer any doubt with respect to the cause of these appearances, being persuaded that the motion in these liquids was occasioned by their particles *going individually*, and *in succession*, to give off their Heat to the cold sides of the tube in the same manner as I have shown in another place that the particles of air give off *their* Heat to other bodies, I was led to conclude that these, and probably all other liquids, are in fact *non-conductors* of

Heat, and I went to work immediately to contrive experiments to put the matter out of all doubt.

On considering the subject attentively, it appeared to me that if liquids were in fact *non-conductors* of Heat, or if it be propagated in them *only* in consequence of the internal motions of their particles, in that case everything which tends to obstruct those motions ought certainly to retard the operation, and render the propagation of the Heat slower and more difficult. I had found that this is actually the case in respect to air, and though (under the influence of a strong and deep-rooted prejudice) I had, from the result of one imperfect experiment, too hastily concluded that it did not take place in regard to water, yet I now found strong reasons to call in question the result of that experiment, and to give the subject a careful and thorough investigation.

Thinking that the best mode of proceeding in this enquiry would be to adopt a method similar to that I had pursued in my experiments on the conducting power of Air, I prepared an apparatus suitable to that purpose. The first object I had in view being to discover whether the propagation of Heat through water was obstructed or not, by rendering the internal motion among the particles of the water, occasioned by their change of temperature, embarrassed and difficult, I contrived to make a certain quantity of Heat pass through a certain quantity of pure water confined in a certain space; and, noting the time employed in this operation, I repeated the experiment again with the same apparatus, with this difference only, that in this second trial the water through which the Heat was made to pass, instead of being pure, was mixed with a small quantity of some fine substance (such as eider-down, for instance), which, without altering any of its chemical properties, or impairing its fluidity, served merely to obstruct and embarrass the motions of the particles of the water in transporting the Heat, in case Heat should be actually *transported* or *carried* in this manner, and not suffered to pass freely through liquids.

The body which received the heat, and which served at the same time to measure the quantity of it communicated, was a very large cylindrical thermometer. (See Plate 4.) The bulb of this

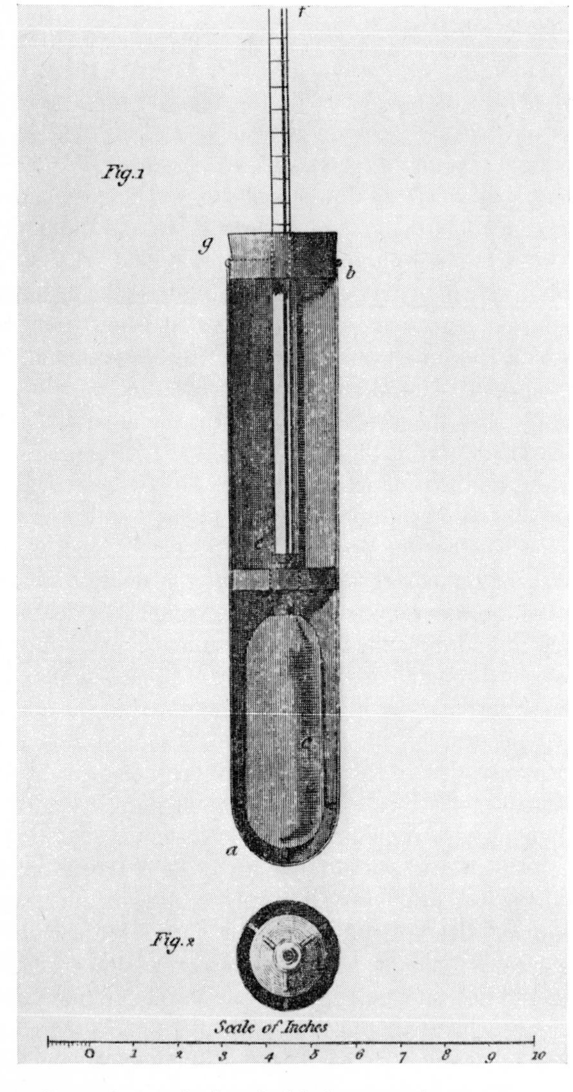

PLATE 4. Rumford's cylindrical passage thermometer.

thermometer, which is constructed of thin sheet-copper, is cylindrical, its two ends being hemispheres.

Its dimensions are as follows:

Dimensions of the bulb of the thermometer.	Diameter	1.84 inches.
	Length	4.99 ,,
	Capacity or contents	13.2099 cubic inches.
	External superficies	28.834 superficial inches.

The thickness of the sheet-copper of which it is constructed is 0.03 of an inch. It weighs, empty, 1846 grains, and is capable of containing 3344 grains of water at the temperature of 55°. This copper bulb has a glass tube, 24 inches long, and $\frac{4}{10}$ of an inch in diameter, which is fitted by means of a good cork into a cylindrical tube or neck of copper, one inch long, and $\frac{65}{100}$ of an inch in diameter, belonging to the metallic bulb.

This thermometer, being filled with linseed-oil and its scale graduated, was fixed in the axis of a hollow cylinder constructed of thin sheet-copper, $11\frac{1}{2}$ inches long, and 2.3435 inches in diameter internally. This cylinder, which is open at one end, is closed at the other with a hemispherical bottom, with its convex surface outwards. The cylinder weighs 2261 grains, and the sheet-brass, of which it is constructed, is 0.0128 of an inch in thickness.

The bulb of the thermometer was placed in the lower part of this brass cylindrical tube, and was confined in the middle or axis of it by means of three pins of wood, about $\frac{1}{10}$ of an inch in diameter, and $\frac{1}{4}$ of an inch long, which pins are fixed in tubes of thin sheet-brass $\frac{1}{10}$ of an inch in diameter, and $\frac{3}{20}$ of an inch in length. These short tubes, which are placed at proper distances on the inside of the large brass tube at its lower end, and firmly attached to it by solder, serve as sockets into which the ends of the wooden pins are fixed, which, pointing inwards or towards the axis of the large cylindrical tube, serve to confine the lower end of the bulb of the thermometer in its proper place. Its upper end is kept in its place, or the axis of the thermometer is made to coincide with the axis of the brass cylinder, by causing the tube of the thermometer to pass through a hole in the middle of a cork stopper which closes the end of the cylinder.

The bottom of the bulb of the thermometer does not repose on the hemispherical bottom of the brass cylinder, but is supported at the distance of $\frac{1}{4}$ of an inch above it, on the end of a wooden pin, like those just described, which pin is fixed in a socket in the middle of the bottom of the cylindrical tube and projects upwards. The ends of all these wooden pins which project beyond the sockets in which they are fixed are reduced to a blunt point. This was done to reduce as much as possible the points of contact between the ends of these pins and the bulb of the thermometer.

The thermometer being in its place, there is on every side a void space left between the bulb of the thermometer and the internal surface of the brass cylinder in which it is confined, the distance between the external surface of the bulb of the thermometer and the internal surface of the containing cylinder being 0.25175 of an inch. This space is designed to contain the water and other substance through which the Heat is made to pass *into* or *out of*, the bulb of the thermometer, and the quantity of Heat which has passed is shewn by the height of the fluid in the tube of the thermometer. The quantity of water required to fill this space and to cover the upper end of the bulb of the thermometer to the height of about $\frac{1}{4}$ of an inch was found to weigh 2468 grains. As the thermometer was plunged into this water, it was, of course, in contact with it by its whole surface, which, as we have seen, is equal to 28.834 square inches.

The bulb of the thermometer being surrounded by water, or by any other liquid or mixture, the conducting power of which was to be ascertained, a cylinder of cork something less in diameter than the brass cylinder, about half an inch long, with a hole in its center, in which the tube of the thermometer passed freely, was thrust down into the brass cylinder, but not quite so low as to touch the surface of the water or other substance it contained. This cylinder, or disk, was supported in its proper place by three projecting brass points or pins which were fixed with solder to the outside of the metallic neck of the bulb of the thermometer.

As soon as this disk of cork is put into its place, the upper part of the hollow brass cylindrical tube is filled with eider-down, and it is closed above with its cork stopper, the tube of the thermometer, which passes through a fit hole in the middle of this stopper, projecting upwards. As the whole scale of the thermometer, from the point of freezing to that of boiling water, is above the upper surface of this stopper, all the changes of Heat to which the instrument is exposed can be observed at all times without deranging any part of the apparatus.

The thermometer is divided according to the scale of Fahrenheit, and its divisions are made to correspond with a very accurate mercurial thermometer made by Troughton.

The experiments with this instrument, which, for the sake of distinction, I shall call my *cylindrical passage thermometer*, were made in the following manner: The thermometer being fixed in its cylindrical brass tube in the manner above described, and surrounded by the substance the conducting power of which was to be ascertained, the instrument was placed in thawing ice, where it was suffered to remain till the thermometer fell to 32°. It was then taken out of the melting ice and immediately plunged into a large vessel of boiling water, and the conducting power of the substance which was the subject of the experiment was estimated by the time employed by the Heat in passing through it into the thermometer; the time being carefully noted when the liquid in the thermometer arrived at the 40th degree of its scale, and also when it came to every 20th degree above it.

As the slower Heat moves, or is transported, in any medium, the longer must of course be the time required for any given quantity of it to pass through it; and as the thermometer shows the changes which take place in the temperature of the body which is heated or cooled (namely, the liquid with which the thermometer is filled), in consequence of the passage of the Heat through the medium by which the thermometer is surrounded, the conducting power of that medium is shewn by the quickness of the ascent or descent of the thermometer, when, having been previously brought to a certain temperature, the instrument is suddenly removed and plunged into another medium at any other constant given temperature.

Having still fresh in my memory the accidents I had so often met with in eating hot apple-pies, I was very impatient, when I had completed this instrument, to see if apples, which, as I well knew, are composed almost entirely of water, really possess a greater power of retaining Heat than that liquid when it is pure or unmixed with other bodies. But before I made the experiment, in order that its result might be the more satisfactory, I determined in the following manner how much water there really is in apples, and what proportion their fibrous parts bear to their whole volume.

960 grains of stewed apples (the apples having been carefully pared and freed from their stems and seeds before they were stewed) were well washed in a large quantity of cold spring water, and the fibrous parts of the apples being suffered to subside to the bottom of the vessel, the clear part of the liquor was poured off, and the fibrous remainder being thoroughly dried was carefully weighed, and was found to weigh just 25 grains.

This fibrous remainder of the 960 grains of stewed apples being again washed in a fresh quantity of cold spring water, and afterwards very thoroughly dried by being exposed several days on a china plate placed on the top of a German stove, which was kept constantly hot, was again weighed, and was found to weigh no more than $18\frac{9}{16}$ grains.

From this experiment it appears that the fibrous parts of stewed apples amount to less than $\frac{1}{50}$ part of the whole mass, and there is abundant reason to conclude that the remainder, amounting to $\frac{49}{50}$ of the whole, is little else than pure water.

Having surrounded the bulb of my cylindrical passage thermometer with a quantity of these stewed apples (the consistence of the mass being such that it shewed no signs of fluidity), the instrument was placed in pounded ice which was melting, and when the thermometer indicated that the whole was cooled down to the temperature of 32°, the instrument was taken out of the melting ice and plunged into a large vessel of boiling water, and the water being kept boiling with the utmost violence during the whole time the experiment lasted, the times taken up in heating the thermometer from 20 to 20 degrees were observed and noted down in a table which had been previously prepared for that purpose.

This experiment having been repeated twice, and varied as often by first heating the instrument to the temperature of boiling water and then plunging it into melting ice, and observing the time taken up in the passage of the Heat *out* of the thermometer, I removed the stewed apples which surrounded the bulb of the thermometer, and, filling the space they had occupied with *pure water*, I now repeated the experiments again with that liquid. The following tables shew the results of these experiments.

The results of these experiments shew that Heat passes with much greater difficulty, or much slower, in *stewed apples* than in *pure water*; and as stewed apples are little else than water mixed with a very small proportion of fibrous and mucilaginous matter, this shews that the conducting power of water with regard to Heat *may be impaired.*

The results of the following experiments will serve to confirm this conclusion.

As the heating or cooling of the instrument goes on very slowly when it approaches to the temperature of the medium in which it is placed, while, on the other hand, this process is very rapid when, the temperature of the instrument being very different from that of the medium, it is first plunged into it, both these circumstances conspire to render the observations made at the extremities of the scale of the thermometer more subject to error, and consequently less satisfactory, than those made nearer the middle of it. In order that the general conclusions drawn from the result of the experiments might not be vitiated by the effects produced by these unavoidable inaccuracies, instead of estimating the celerity of the passage of the Heat by the times elapsed in heating and cooling the thermometer *through the whole length of its scale*, or between the point of freezing to that of boiling water, I have taken the times elapsed in heating and cooling it 80 *degrees in the middle of the scale*, *viz.* between 80° and 160°, as the measure of the conducting powers of the substances through which the Heat was made to pass.

TABLE 1

	Time the Heat was passing INTO the Thermometer			
	Through Stewed Apples		Through Water	
	Exp. No. 1	Exp. No. 3	Exp. No. 5	Exp. No. 7
	Seconds	Seconds	Seconds	Seconds
In heating the Thermometer from the temperature of 32° to that of 40	95	89	45	45
from 40° to 60	75	67	36	35
60 to 80	61	56	34	31
100	65	60	30	30
120	73	66	37	36
140	90	82	44	44
160	121	113	63	60
180	188	170	93	90
200	360	364	226	215
Total times in heating from 32° to 200°	1128	1057	608	586
Times employed in heating the instrument 80 degrees, *viz.* from 80° to 160° .	349″	321″	174″	170″
Mean times in heating it from 80° to 160° . . .	In Stewed Apples 335″		In Water 172″	

I have, however, noted the times which elapsed in heating and cooling the instrument through a much larger interval, namely, through an interval of 168 degrees in *heating*, or from 32° to 200°, and in *cooling* through 160 degrees, or from 200° to 40°.

In respect to the *cooling* of the instrument, it is necessary that I should inform my reader, that, though I have not in the tables of the experiments mentioned any higher temperature than that of 200°, yet the instrument was always heated to the point of boiling water, which, under the pressure of the atmosphere at Munich, where the experiments were made, was commonly about 209½ deg. of Fahrenheit's scale. The instrument, being kept in boiling water till its thermometer appeared to be quite stationary, was then taken out of the water and instantly plunged into melting ice, and the time was observed and carefully noted down when the liquid in its thermometer passed the division of its scale which indicated 200°, as also when it arrived at the other divisions indicated in the tables.

With regard to the four last-mentioned experiments (No. 2, 4, 6, and 8), it will be found, on examination, that their results

TABLE 2

	Time the Heat was passing OUT of the Thermometer			
	Through Stewed Apples		Through Water	
	Exp. No. 2	Exp. No. 4	Exp. No. 6	Exp. No. 8
	Seconds	Seconds	Seconds	Seconds
In cooling the Thermometer from the temperature of 200° to that of . . . 180	80	74	46	37
from 180° to 160°	75	72	42	37
160 to 140	84	83	43	43
120	107	101	54	51
100	141	136	73	73
80	198	190	112	105
60	321	307	200	204
40	775	733	483	461
Total time in cooling from 200° to 40°	1781	1696	1053	1011
Times employed in cooling the instrument 80 degrees, *viz.* from 160° to 80° .	530″	510″	282″	272″
Mean time in cooling it from 160° to 80°	In Stewed Apples 520″		In Water 277″	

correspond very exactly with those before described; and they certainly prove in a very decisive manner this important fact,—*that a small proportion of certain substances, on being mixed with water, tend very powerfully to impair the conducting power of that Fluid in regard to Heat.*

In the experiments No. 1 and No. 2, which were both made on the same day, and in the order in which they are numbered, the Heat was considerably more obstructed in its passage through the mass of stewed apples which surrounded the thermometer than in the experiments No. 3 and No. 4, which were made on the following day. It is probable that this was occasioned by some change in the consistency of this soft mass of the stewed apples which had taken place while the instrument was left to repose in the interval between the experiments; but instead of stopping to show how this might be explained, I shall proceed to give an account of some experiments from the results of which we shall derive information that will be much more satisfactory than any speculations I could offer on that subject.

Supposing Heat to be propagated in water in the same manner as it is propagated in air and other elastic Fluids, namely, that it is *transported* by its particles, these particles being put in motion by the change which is produced in their specific gravity by the change of temperature, and that there is no communication whatever, or *interchange of their Heat*, among the particles of *the same Fluid*; in that case it is evident that the propagation of Heat in a Fluid may be obstructed in two ways, namely, by diminishing its *Fluidity* (which may be done by dissolving in it any mucilaginous substance); or, more simply, by merely embarrassing and obstructing the motion of its particles in the operation of transporting the Heat, which may be effected by mixing with the Fluid any solid substance (it must be a non-conductor of Heat, however) in small masses, or which has a very large surface in proportion to its solidity.

In the foregoing experiments with *stewed apples*, the passage of the Heat in the water (which constituted by far the greatest part of the mass) was doubtless obstructed in both these ways. The mucilaginous parts of the apples diminished very much the fluidity of the water, at the same time that the fibrous parts served to embarrass its internal motions.

In order to discover the *comparative effects* of these two causes, it was necessary to separate them, or to contrive experiments in which only one of them should be permitted to act at the same time. This I endeavoured to do in the following manner.

To ascertain the effects produced by diminishing the *fluidity* of water, I mixed with it a small quantity of starch, namely, 192 grains in weight to 2276 grains of water; and to determine the effects produced by merely *embarrassing* the water in its motions, I mixed with it an equal proportion (by weight) of *eider-down*. The starch was boiled with the water with which it was mixed, as was also the eider-down. This last-mentioned substance was boiled in the water in order to free it from air, which, as is well known, adheres to it with great obstinacy.

In order that these experiments may with greater facility be compared with those which were made with *stewed apples* and with *pure water*, I shall place their results all together in the following tables.

As the results of these experiments prove, in the most decisive manner, that the propagation of Heat in water is retarded, not only by those things which diminish its fluidity, but also by those which, by mechanical means, and without forming any combination with it whatever, merely obstruct its internal motions, it appears to me that this proves, almost to a demonstration, that Heat is propagated in water *in consequence* of its internal motions, or that it is transported or *carried* by the particles of that liquid, and that it does not spread and expand in it, as has generally been imagined.

I have shewn in another place, and I believe I may venture to say I have proved,* that Heat is actually propagated in *air* in the same manner I here suppose it to be propagated in water, and if the conducting powers of both these fluids are found to be impaired by the *same means*, it affords very strong grounds to

* See Philosophical Transactions, 1792.

conclude that they both conduct Heat in the *same manner*; but this has been found to be actually the case.

Eider-down, which cannot affect the specific qualities of either of those fluids, and which certainly does no more when mixed with them than merely to obstruct and embarrass their internal motions, has been found to retard very much the propagation of Heat in both of them: on comparing these experiments with those I formerly made on the conducting power of air, it will even be found that the conducting power of water is nearly, if not quite, as much impaired by a mixture of *eider-down* as that of air.

In the course of my experiments on the various substances used in forming artificial clothing for confining Heat, I found that the thickness of a stratum of air, which served as a barrier to Heat, remaining the same, the passage of Heat through it was sometimes rendered more difficult by increasing the quantity of the light substance which was mixed with it to obstruct its internal motion.

To see if similar effects would be produced by the same means when Heat is made to pass through water, I repeated the experiments with *eider-down*, reducing the quantity of it mixed with the water to 48 grains, or *one quarter* of the quantity used in the experiments No. 11 and No. 12.

The results of these experiments, and a comparison of them with those before mentioned, may be seen in the following tables.

The results of these experiments are extremely interesting. They not only make us acquainted with a new and very curious fact, namely, that feathers and other like substances, which, in air, are known to form very warm covering for confining Heat, not only serve the same purpose in water, but that their effect in preventing the passage of Heat is even greater in water than in air.

This discovery, if I do not deceive myself, throws a very broad light over some of the most interesting parts of the economy of Nature, and gives us much satisfactory information respecting the final causes of many phænomena which have hitherto been little understood.

As *liquid water* is the vehicle of Heat and nourishment, and consequently of life, in every living thing; and as water, left to

TABLE 3

	Time the Heat was in passing INTO the Thermometer			
	Through a Mixture of 2276 Grains of Water and 192 Grains of Starch	Through a Mixture of 2276 Grains of Water and 192 Grains of Eider-Down	Through Stewed Apples	Through Pure Water
	Experiment No. 9	Experiment No. 11	Mean of Two Experiments No. 1 and No. 3	Mean of Two Experiments No. 5 and No. 7
	Seconds	Seconds	Seconds	Seconds
In heating the Thermometer from 32° to 40°	101	83	92	45
from 40 to 60	72	55	71	35½
60 to 80	64	49	58½	32½
100	63	52	62½	30
120	74	57	69½	36½
140	89	67	86	44
160	115	93	117	61½
180	178	133	179	91½
200	453	360	362	220½
Total times in heating the instrument from 32° to 200°	1209	949	1096½	597
Times employed in heating the Thermometer 80 degrees, *viz.* from 80° to 160°	341″	269″	335″	172″

itself, freezes, with a degree of cold much less than that which frequently prevails in cold climates, it is agreeable to the ideas we have of the wisdom of the Creator of the world to expect that effectual measures would be taken to preserve a sufficient quantity of that liquid in its fluid state to maintain life during the cold season: and this we find has actually been done; for both plants and animals are found to survive the longest and most severe winters; but the means which have been employed to produce this

TABLE 4

	Time the Heat was in passing OUT OF the Thermometer			
	Through a Mixture of 2276 Grains of Water and 192 Grains of Starch	Through a Mixture of 2276 Grains of Water and 192 Grains of Eider-Down	Through Stewed Apples	Through Pure Water
	Experiment No. 10	Experiment No. 12	Mean of Two Experiments No. 2 and No. 4	Mean of Two Experiments No. 6 and No. 8
	Seconds	Seconds	Seconds	Seconds
In cooling the Thermometer from 200° to 180° . . .	69	68	77	41½
from 180° to 160°	66	61	73½	39½
160 to 140	74	72	83½	43
120	92	91	104	52½
100	119	120	138½	73
80	173	177	194	108½
60	283	279	314	202
40	672	673	754	472
Total times in cooling from 200° to 40°	1548	1541	1749½	1032
Times employed in cooling the instrument 80 degrees, *viz.* from 160° to 80°	468″	460″	520″	277″

TABLE 5

	Time the Heat was in passing INTO the Thermometer		
	Through Water with 48 Grains, or $\frac{1}{50}$ of its Bulk of Eider-Down	Through Water with 192 Grains, or $\frac{4}{50}$ of its Bulk of Eider-Down	Through Pure Water
	Experiment No. 13	Experiment No. 11	Mean of Two Experiments, No. 5 and No. 7
	Seconds	Seconds	Seconds
In heating the Thermometer from 32° to 40°	51	83	45
from 40 to 60	47	55	35½
60 to 80	39	49	32½
100	40	52	30
120	45	57	36½
140	56	67	44
160	74	93	61½
180	118	133	91½
200	293	360	220½
Total times in heating from 32° to 200° . .	763	949	597
Times employed in heating the instrument 80 degrees, or from 80° to 160°	215″	269″	172″

miraculous effect have not been investigated,—at least not in as far as they relate to vegetables.

But as animal and vegetable bodies are essentially different in many respects, it is very natural to suppose that the means would be different which are employed to preserve them against the fatal effects which would be produced in each by the congelation of their fluids.

Among organized bodies which live on the surface of the earth, and which, of course, are exposed to the vicissitudes of the seasons, we find that as the proportion of fluids to solids is greater, the greater is the Heat which is required for the support of life and health, and the less are they able to endure any considerable change of their temperature.

The proportion of Fluids to Solids is much greater in *animals* than in vegetables; and in order to preserve in them the great

TABLE 6

	Time the Heat was passing OUT OF the Thermometer		
	Through Water with 48 Grains, or $\frac{1}{50}$ of its Bulk of Eider-Down	Through Water with 192 Grains, or $\frac{4}{50}$ of its Bulk of Eider-Down	Through Pure Water
	Experiment No. 14	Experiment No. 12	Mean of Two Experiments, No. 6 and No. 8
	Seconds	Seconds	Seconds
In cooling the Thermometer from 200° to 180°	49	68	41½
from 180 to 160	50	61	39½
160 to 140	56	72	43
120	70	91	52½
100	96	120	73
80	151	177	108½
60	262	279	202
40	661	673	472
Total times in cooling from 200° to 40° . .	1395	1541	1032
Times employed in cooling the instrument 80 degrees, *viz.* from 160° to 80°	373″	460″	277″

quantity of Heat which is necessary to the preservation of life, they are furnished with lungs, and are warmed by a process similar to that by which Heat is generated in the combustion of inflammable bodies.

Among *vegetables,* those which are the most succulent are *annual.* Not being furnished with lungs to keep the great mass of liquids warm, which fill their large and slender vessels, they live only while the genial influence of the sun warms them and animates their feeble powers, and they droop and die as soon as they are deprived of his support.

There are many tender plants to be found in cold countries, which die in the autumn, the roots of which remain alive during the winter, and send off fresh shoots in the ensuing spring. In these we shall constantly find the roots more compact and dense than the stalk, or with smaller vessels and a smaller proportion of Fluids.

Among the trees of the forest we shall constantly find that those which contain a great proportion of *thin watery liquids* not only shed their leaves every autumn, but are sometimes frozen, and actually killed, in severe frosts. Many thousands of the largest walnut-trees were killed by the frost in the Palatinate during the very cold winter in the year 1788; and it is well known that few, if any, of the deciduous plants of our temperate climate would be able to support the excessive cold of the frigid zone.

The trees which grow in those inhospitable climates, and which brave the colds of the severest winters, contain very little watery liquids. The sap which circulates in their vessels is thick and viscous, and can hardly be said to be fluid. Is there not the strongest reason to think that this was so contrived for the express purpose of preventing their being deprived of all their Heat, and killed by the cold during the winter?

We have seen by the foregoing experiments how much the propagation of Heat in a liquid is retarded by diminishing its fluidity; and who knows but this may continue to be the case as long as any degree of fluidity remains?

As the bodies and branches of trees are not covered in winter by the snow which protects their roots from the cold atmosphere, it is evident that extraordinary measures were necessary to prevent their being frozen. The bark of all such trees as are designed by nature to support great degrees of cold forms a very warm covering; but this precaution alone would certainly not have been sufficient for their protection. The sap in all trees which are capable of supporting a long continuance of frost grows thick and viscous on the approach of winter. What more important purpose could this change answer than that here indicated? And it would be more than folly to pretend that it answers no useful purpose at all.

We have seen by the results of the foregoing experiments how much the simple embarrassment of liquids in their internal motions tends to retard the propagation of Heat in them, and consequently its passage out of them;—and when we consider the extreme smallness of the vessels in which the sap moves in vegetables and particularly in large trees; when we recollect that the substance of which these small tubes are formed is one of the best non-conductors of heat known;* and when we advert to the additional embarrassments to the passage of the Heat which arise from the increased viscosity of the sap in winter, and to the almost impenetrable covering for confining Heat which is formed by the bark, we shall no longer be at a loss to account for the preservation of trees during the winter, notwithstanding the long continuation of the hard frosts to which they are annually exposed.

* I lately, by accident, had occasion to observe a very striking proof of the extreme difficulty with which Heat passes in wood. Being present at the foundry at Munich when cannons were casting, I observed that the founder used a wooden instrument for stirring the melted metal. It was a piece of oak plank, green or unseasoned, about ten inches square and two inches thick, with a long wooden handle which was fitted into a hole in the middle of it. As this instrument was frequently used, and sometimes remained a considerable time in the furnace, in which the Heat was most intense, I was surprised to find that it was not consumed; but I was still more surprised, on examining the part of the plank which had been immersed in the melted metal, to find that the Heat had penetrated it to so inconsiderable a depth, that, at the distance of one twentieth of an inch below its surface, the wood did not seem to have been in the least affected by it. The colour of the wood remained unchanged, and it did not appear to have lost even its moisture.

On the same principles we may, I think, account in a satisfactory manner for the preservation of several kinds of fruit—such as apples and pears, for instance—which are known to support, without freezing, a degree of cold which would soon reduce an equal volume of *pure water* to a solid mass of ice.

At the same time that the compact skin of the fruit effectually prevents the evaporation of its fluid parts, which, as is well known, could not take place without occasioning a very great loss of Heat, the internal motions of those fluids are so much obstructed by the thin partitions of the innumerable small cells in which they are confined, that the communication of their Heat to the air ought, according to our hypothesis, to be extremely slow and difficult. These fruits do, however, freeze at last, when the cold is very intense; but it must be remembered that they are composed almost entirely of liquids, and of such liquids as do not grow viscous with cold, and, moreover, that they were evidently not designed to support for a long time very severe frosts.

Parsnips and carrots, and several other kinds of roots, support cold without freezing still longer than apples and pears, but these are less watery, and I believe the vessels in which their fluids are contained are smaller; and both these circumstances ought, according to our assumed principles, to render the passage of their Heat out of them more difficult, and consequently to retard their congelation.

But there is still another circumstance, and a very remarkable one indeed, which, if our conjectures respecting the manner in which Heat is propagated in liquids be true, must act a most important part in the preservation of Heat, and consequently of animal and vegetable life, in cold climates. But as the probability of all these deductions must depend very much on the evidence which is brought to prove the great fundamental fact on which they are established,—that respecting the internal motions among the particles of liquids which *necessarily* take place when they are heated or cooled,—before I proceed any farther in these speculations, I shall endeavour to throw some more light on that curious and interesting subject.

CHAPTER 4

HEAT BY FRICTION

AMONG the most famous experiments conducted for the purpose of disproving the materiality of heat were those of Count Rumford, described by him in the following pages. We have already seen in Chapter 1 the place this experiment enjoys in the historical perspective of the development of our modern theory of heat.

Rumford's name is often associated with that of his protégé, Humphrey Davy, in connection with experiments of heat produced by friction. A study of Davy's experiments, which were carried out by Davy in 1798 when he was a lad of 19, show that the value of his contribution has been grossly overrated.† They were published in *Contributions to Physical and Medical Knowledge Principally from the West of England* which was edited by Davy's employer, Dr. Thomas Beddoes. Even though these experiments cannot be taken seriously by any physicist who cares to look into the actual experiments, they did have the very important result of calling Rumford's attention to Davy and starting him on his road to fame.

ESSAY IX

An Experimental Inquiry concerning the Source of the Heat which is Excited by Friction

Read before the Royal Society, January 25, 1798

Philosophical Transactions LXXXVIII, 80–102 (1798)
Bibliotheque Britannique VIII, 3–34 (1798)
Nicholson's quarto Journal I, 459–468, 515–518 (1798)
Gilbert's Annalen der Physik IV, 257–281, 377–399 (1798)
Scherer's Journal der Chemie I, 9–31 (1798)
Voigt's Magazin I, 94–106 (1798)

† E. N. daC. Andrade, *Nature*, March 9, 1935, p. 359.

It frequently happens that in the ordinary affairs and occupations of life, opportunities present themselves of contemplating some of the most curious operations of Nature; and very interesting philosophical experiments might often be made, almost without trouble or expense, by means of machinery contrived for the mere mechanical purposes of the arts and manufactures.

I have frequently had occasion to make this observation; and am persuaded that a habit of keeping the eyes open to everything that is going on in the ordinary course of the business of life has oftener led, as it were by accident, or in the playful excursions of the imagination, put into action by contemplating the most common appearances, to useful doubts and sensible schemes for investigation and improvement, than all the more intense meditations of philosophers in the hours expressly set apart for study.

It was by accident that I was led to make the experiments of which I am about to give an account; and, though they are not perhaps of sufficient importance to merit so formal an introduction, I cannot help flattering myself that they will be thought curious in several respects, and worthy of the honour of being made known to the Royal Society.

Being engaged lately in superintending the boring of cannon in the workshops of the military arsenal at Munich, I was struck with the very considerable degree of Heat which a brass gun acquires in a short time in being bored, and with the still more intense Heat (much greater than that of boiling water, as I found by experiment) of the metallic chips separated from it by the borer.

The more I meditated on these phænomena, the more they appeared to me to be curious and interesting. A thorough investigation of them seemed even to bid fair to give a farther insight into the hidden nature of Heat; and to enable us to form some reasonable conjectures respecting the existence, or non-existence, of an *igneous fluid*,—a subject on which the opinions of philosophers have in all ages been much divided.

In order that the Society may have clear and distinct ideas of the speculations and reasonings to which these appearances gave rise in my mind, and also of the specific objects of philosophical investigation they suggested to me, I must beg leave to state them at some length, and in such manner as I shall think best suited to answer this purpose.

From *whence comes* the Heat actually produced in the mechanical operation above mentioned?

Is it furnished by the metallic chips which are separated by the borer from the solid mass of metal?

If this were the case, then, according to the modern doctrines of latent Heat, and of caloric, the *capacity for Heat* of the parts of the metal, so reduced to chips, ought not only to be changed, but the change undergone by them should be sufficiently great to account for *all* the Heat produced.

But no such change had taken place; for I found, upon taking equal quantities, by weight, of these chips, and of thin slips of the same block of metal separated by means of a fine saw, and putting them at the same temperature (that of boiling water) into equal quantities of cold water (that is to say, at the temperature of 59½°F.), the portion of water into which the chips were put was not, to all appearance, heated either less or more than the other portion in which the slips of metal were put.

This experiment being repeated several times, the results were always so nearly the same that I could not determine whether any, or what change had been produced in the metal, *in regard to its capacity for Heat*, by being reduced to chips by the borer.*

* As these experiments are important, it may perhaps be agreeable to the Society to be made acquainted with them in their details.

One of them was as follows:

To 4590 grains of water, at the temperature of 59½F. (an allowance as compensation, reckoned in water, for the capacity for Heat of the containing cylindrical tin vessel being included), were added 1016⅛ grains of gun-metal in thin slips, separated from the gun by means of a fine saw, being at the temperature of 210°F. When they had remained together 1 minute, and had been well stirred about, by means of a small rod of light wood, the Heat of the mixture was found to be = 63°.

From this experiment the *specific Heat* of the metal, calculated according to the rule given by Dr. Crawford, turns out to be = 0.1100, that of water being = 1.0000.

An experiment was afterwards made with the metallic chips as follows:

From hence it is evident that the Heat produced could not possibly have been furnished at the expence of the latent Heat of the metallic chips. But, not being willing to rest satisfied with these trials, however conclusive they appeared to me to be, I had recourse to the following still more decisive experiment.

Taking a cannon (a brass six-pounder), cast solid, and rough as it came from the foundry (see Fig. 1 [Plate 5]), and fixing it (horizontally) in the machine used for boring, and at the same time finishing the outside of the cannon by turning (see Fig. 2 [Plate 5]), I caused its extremity to be cut off, and, by turning down the metal in that part, a solid cylinder was formed, $7\frac{3}{4}$ inches in diameter, and $9\frac{8}{10}$ inches long, which, when finished, remained joined to the rest of the metal (that which, properly speaking, constituted the cannon) by a small cylindrical neck, only $2\frac{1}{5}$ inches in diameter, and $3\frac{8}{10}$ inches long.

This short cylinder, which was supported in its horizontal position and turned round its axis by means of the neck by which it remained united to the cannon, was now bored with the horizontal borer used in boring cannon; but its bore, which was 3.7 inches in diameter, instead of being continued through its whole length (9.8 inches) was only 7.2 inches in length; so that a solid bottom was left to this hollow cylinder, which bottom was 2.6 inches in thickness.

This cavity is represented by dotted lines in Fig. 2 [Plate 5]; as also in Fig. 3 [Plate 5], where the cylinder is represented on an enlarged scale.

This cylinder being designed for the express purpose of generating Heat *by friction*, by having a blunt borer forced against its solid bottom at the same time that it should be turned round its

To the same quantity of water as was used in the experiment above mentioned, at the same temperature (*viz.* $59\frac{1}{2}°$), and in the same cylindrical tin vessel, were now put $1016\frac{1}{8}$ grains of metallic chips of gun-metal bored out of the same gun from which the slips used in the foregoing experiment were taken, and at the same temperature (210°). The Heat of the mixture at the end of 1 minute was just 63°, as before; consequently the specific Heat of these metallic chips was = 0.1100. Each of the above experiments was repeated three times, and always with nearly the same results.

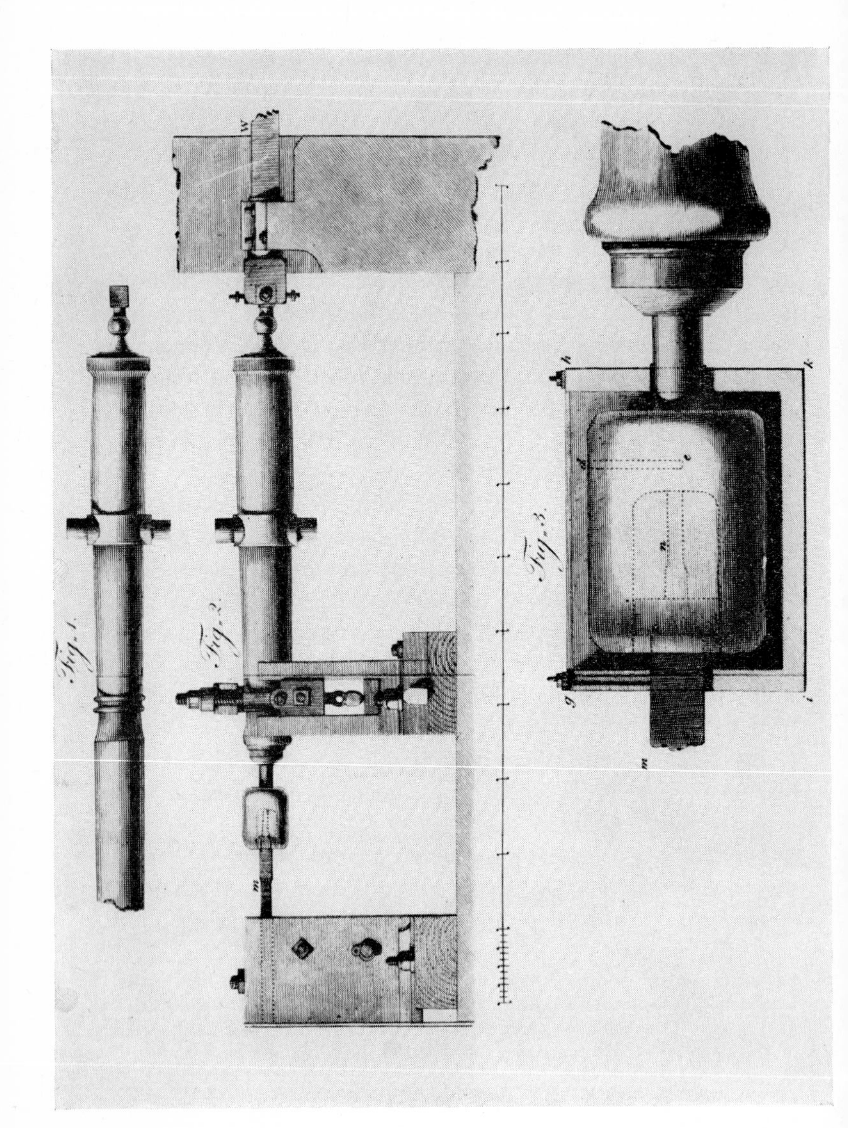

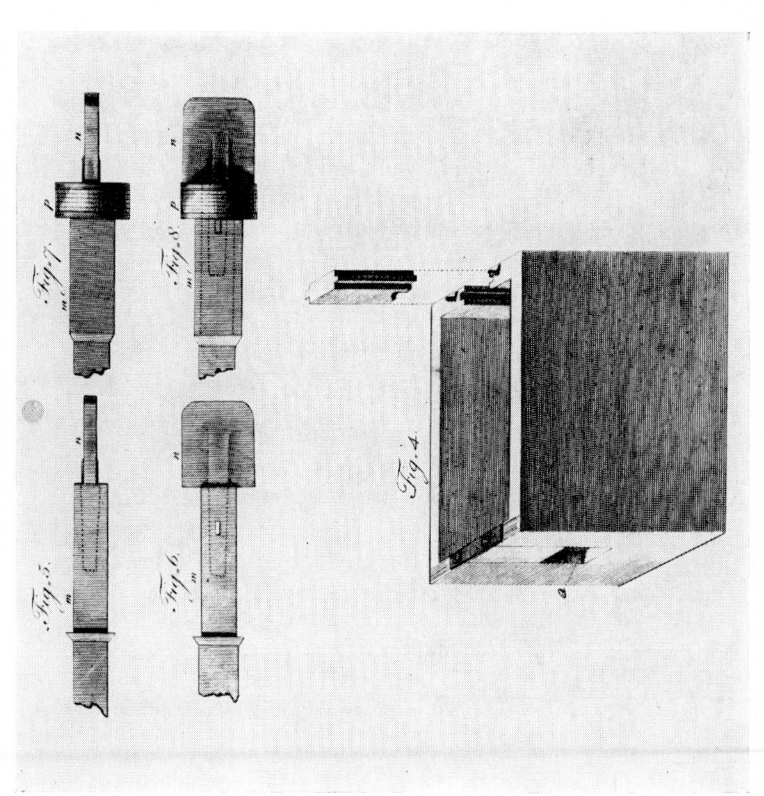

PLATES 5 and 6. Figures 1 to 8 are Rumford's own figures with his original designations.

axis by the force of horses, in order that the Heat accumulated in the cylinder might from time to time be measured, a small round hole (see *d*, *e*, Fig. 3 [Plate 5]), 0.37 of an inch only in diameter, and 4.2 inches in depth, for the purpose of introducing a small cylindrical mercurial thermometer, was made in it, on one side, in a direction perpendicular to the axis of the cylinder, and ending in the middle of the solid part of the metal which formed the bottom of its bore.

The solid contents of this hollow cylinder, exclusive of the cylindrical neck by which it remained united to the cannon, were $385\frac{3}{4}$ cubic inches, English measure, and it weighed 113.13 lb., avoirdupois; as I found on weighing it at the end of the course of experiments made with it, and after it had been separated from the cannon with which, during the experiments, it remained connected.*

Experiment No. 1

This experiment was made in order to ascertain how much Heat was actually generated by friction, when a blunt steel borer being so forcibly shoved (by means of a strong screw) against the bottom of the bore of the cylinder, that the pressure against it was equal to the weight of about 10,000 lb., avoirdupois, the cylinder was turned round on its axis (by the force of horses) at the rate of about 32 times in a minute.

This machinery, as it was put together for the experiment, is

* For fear I should be suspected of prodigality in the prosecution of my philosophical researches, I think it necessary to inform the Society that the cannon I made use of in this experiment was not sacrificed to it. The short hollow cylinder which was formed at the end of it was turned out of a cylindrical base of metal, about 2 feet in length, projecting beyond the muzzle of the gun, called in the German language the *verlorner kopf* (the head of the cannon to be thrown away), and which is represented in Fig. 1 [Plate 5].

This original projection, which is cut off before the gun is bored, is always cast with it, in order that, by means of the pressure of its weight on the metal in the lower part of the mould during the time it is cooling, the gun may be the more compact in the neighbourhood of the muzzle, where, without this precaution, the metal would be apt to be porous, or full of honeycombs.

represented by Fig. 2 [Plate 5]. W is a strong horizontal iron bar, connected with proper machinery carried round by horses, by means of which the cannon was made to turn round its axis.

To prevent, as far as possible, the loss of any part of the Heat that was generated in the experiment, the cylinder was well covered up with a fit coating of thick and warm flannel, which was carefully wrapped round it, and defended it on every side from the cold air of the atmosphere. This covering is not represented in the drawing of the apparatus, Fig. 2 [Plate 5].

I ought to mention that the borer was a flat piece of hardened steel, 0.63 of an inch thick, 4 inches long, and nearly as wide as the cavity of the bore of the cylinder, namely, $3\frac{1}{2}$ inches. Its corners were rounded off at its end, so as to make it fit the hollow bottom of the bore; and it was firmly fastened to the iron bar (*m*) which kept it in its place. The area of the surface by which its end was in contact with the bottom of the bore of the cylinder was nearly $2\frac{1}{3}$ inches. This borer, which is distinguished by the letter *n*, is represented in most of the figures.

At the beginning of the experiment, the temperature of the air in the shade, as also that of the cylinder, was just 60°F.

At the end of 30 minutes, when the cylinder had made 960 revolutions about its axis, the horses being stopped, a cylindrical mercurial thermometer, whose bulb was $\frac{32}{100}$ of an inch in diameter, and $3\frac{1}{4}$ inches in length, was introduced into the hole made to receive it, in the side of the cylinder, when the mercury rose almost instantly to 130°.

Though the Heat could not be supposed to be quite equally distributed in every part of the cylinder, yet, as the length of the bulb of the thermometer was such that it extended from the axis of the cylinder to near its surface, the Heat indicated by it could not be very different from that of the *mean temperature* of the cylinder; and it was on this account that a thermometer of that particular form was chosen for this experiment.

To see how fast the Heat escaped out of the cylinder (in order to be able to make a probable conjecture respecting the quantity given off by it during the time the Heat generated by the friction was accumulating), the machinery standing still, I suffered the thermometer to remain in its place near three quarters of an hour, observing and noting down, at small intervals of time, the height of the temperature indicated by it.

	The Heat, as shown by the thermometer, was
Thus at the end of 4 minutes	126°
after 5 minutes, always reckoning from the first observation	125
at the end of 7 minutes	123
12 ,,	120
14 ,,	119
16 ,,	118
20 ,,	116
24 ,,	115
28 ,,	114
31 ,,	113
34 ,,	112
37½ ,,	111
and when 41 minutes had elapsed	110

Having taken away the borer, I now removed the metallic dust, or, rather, scaly matter, which had been detached from the bottom of the cylinder by the blunt steel borer, in this experiment; and, having carefully weighed it, I found its weight to be 837 grains, Troy.

Is it possible that the very considerable quantity of Heat that was produced in this experiment (a quantity which actually raised the temperature of above 113 lb. of gun-metal at least 70 degrees of Fahrenheit's thermometer, and which, of course, would have been capable of melting 6½ lb. of ice, or of causing near 5 lb. of ice-cold water to boil) could have been furnished by so inconsiderable a quantity of metallic dust? and this merely in consequence of *a change* of its capacity for Heat?

As the weight of this dust (837 grains, Troy) amounted to no more than $\frac{1}{948}$ part of that of the cylinder, it must have lost no less than 948 degrees of Heat, to have been able to have raised the temperature of the cylinder 1 degree; and consequently it must have given off 66,360 degrees of Heat to have produced the effects which were actually found to have been produced in the experiment!

But without insisting on the improbability of this supposition, we have only to recollect, that from the results of actual and decisive experiments, made for the express purpose of ascertaining that fact, the capacity for Heat of the metal of which great guns are cast *is not sensibly changed* by being reduced to the form of metallic chips in the operation of boring cannon; and there does not seem to be any reason to think that it can be much changed, if it be changed at all, in being reduced to much smaller pieces by means of a borer that is less sharp.

If the Heat, or any considerable part of it, were produced in consequence of a change in the capacity for Heat of a part of the metal of the cylinder, as such change could only be *superficial*, the cylinder would by degrees be *exhausted*; or the quantities of Heat produced in any given short space of time would be found to diminish gradually in successive experiments. To find out if this really happened or not, I repeated the last-mentioned experiment several times with the utmost care; but I did not discover the smallest sign of exhaustion in the metal, notwithstanding the large quantities of Heat actually given off.

Finding so much reason to conclude that the Heat generated in these experiments, or *excited*, as I would rather choose to express it, was not furnished *at the expense of the latent Heat* or *combined caloric* of the metal, I pushed my inquiries a step farther, and endeavoured to find out whether the air did, or did not, contribute anything in the generation of it.

Experiment No. 2

As the bore of the cylinder was cylindrical, and as the iron bar (*m*), to the end of which the blunt steel borer was fixed, was square, the air had free access to the inside of the bore, and even to the bottom of it, where the friction took place by which the Heat was excited.

As neither the metallic chips produced in the ordinary course of the operation of boring brass cannon, nor the finer scaly particles produced in the last-mentioned experiments by the friction of the blunt borer, showed any signs of calcination, I did not see how the air could possibly have been the cause of the heat that was produced; but, in an investigation of this kind, I thought that no pains should be spared to clear away the rubbish, and leave the subject as naked and open to inspection as possible.

In order, by one decisive experiment, to determine whether the air of the atmosphere had any part, or not, in the generation of the Heat, I contrived to repeat the experiment under circumstances in which *it was evidently impossible for it to produce any effect whatever*. By means of a piston exactly fitted to the mouth of the bore of the cylinder, through the middle of which piston the square iron bar, to the end of which the blunt steel borer was fixed, passed in a square hole made perfectly air-tight, the access of the external air to the inside of the bore of the cylinder was effectually prevented. (In Fig. 3 [Plate 5], this piston (*p*) is seen in its place; it is likewise shown in Fig. 7 and 8 [Plate 6].)

I did not find, however, by this experiment, that the exclusion of the air diminished, in the smallest degree, the quantity of Heat excited by the friction.

There still remained one doubt, which, though it appeared to me to be so slight as hardly to deserve any attention, I was however desirous to remove. The piston which closed the mouth of the bore of the cylinder, in order that it might be air-tight, was fitted into it with so much nicety, by means of its collars of leather, and pressed against it with so much force, that, notwithstanding its being oiled, it occasioned a considerable degree of friction when the hollow cylinder was turned round its axis. Was not the Heat produced, or at least some part of it, occasioned by this friction of the piston? and, as the external air had free access to the extremity of the bore, where it came in contact with the piston, is it not possible that this air may have had some share in the generation of the Heat produced?

Experiment No. 3

A quadrangular oblong deal box (see Fig. 4 [Plate 6]), water-tight, $11\frac{1}{2}$ English inches long, $9\frac{4}{10}$ inches wide, and $9\frac{6}{10}$ inches deep (measured in the clear), being provided with holes or slits in the middle of each of its ends, just large enough to receive, the one the square iron rod to the end of which the blunt steel borer was fastened, the other the small cylindrical neck which joined the hollow cylinder to the cannon; when this box (which was occasionally closed above by a wooden cover or lid moving on hinges) was put into its place, that is to say, when, by means of the two vertical openings or slits in its two ends (the upper parts of which openings were occasionally closed by means of narrow pieces of wood sliding in vertical grooves), the box (*g*, *h*, *i*, *k*, Fig. 3 [Plate 5]) was fixed to the machinery in such a manner that its bottom (*i*, *k*) being in the plane of the horizon, its axis coincided with the axis of the hollow metallic cylinder; it is evident, from the description, that the hollow metallic cylinder would occupy the middle of the box, without touching it on either side (as it is represented in Fig. 3 [Plate 5]); and that, on pouring water into the box, and filling it to the brim, the cylinder would be completely covered and surrounded on every side by that fluid. And farther, as the box was held fast by the strong square iron rod (*m*) which passed in a *square hole* in the center of one of its ends (*a*, Fig. 4 [Plate 6]), while the round or cylindrical neck, which joined the hollow cylinder to the end of the cannon, could turn round freely on its axis in the *round hole* in the center of the other end of it, it is evident that the machinery could be put in motion without the least danger of forcing the box out of its place, throwing the water out of it, or deranging any part of the apparatus.

Everything being ready, I proceeded to make the experiment I had projected in the following manner.

The hollow cylinder having been previously cleaned out, and the inside of its bore wiped with a clean towel till it was quite dry, the square iron bar, with the blunt steel borer fixed to the end of it, was put into its place; the mouth of the bore of the cylinder being closed at the same time by means of the circular piston, through the center of which the iron bar passed.

This being done, the box was put in its place, and the joinings of the iron rod and of the neck of the cylinder with the two ends of the box having been made watertight by means of collars of oiled leather, the box was filled with cold water (*viz.* at the temperature of 60°), and the machine was put in motion.

The result of this beautiful experiment was very striking, and the pleasure it afforded me amply repaid me for all the trouble I had had in contriving and arranging the complicated machinery used in making it.

The cylinder, revolving at the rate of about 32 times in a minute, had been in motion but a short time, when I perceived, by putting my hand into the water and touching the outside of the cylinder, that Heat was generated; and it was not long before the water which surrounded the cylinder began to be sensibly warm.

At the end of 1 hour I found, by plunging a thermometer into the water in the box (the quantity of which fluid amounted to 18.77 lb., avoirdupois, or $2\frac{1}{4}$ wine gallons), that its temperature had been raised no less than 47 degrees; being now 107° of Fahrenheit's scale.

When 30 minutes more had elapsed, or 1 hour and 30 minutes after the machinery had been put in motion, the Heat of the water in the box was 142°.

At the end of 2 hours, reckoning from the beginning of the experiment, the temperature of the water was found to be raised to 178°.

At 2 hours 20 minutes it was at 200°; and at 2 hours 30 minutes it ACTUALLY BOILED!

It would be difficult to describe the surprise and astonishment expressed in the countenances of the bystanders, on seeing so large a quantity of cold water heated, and actually made to boil, without any fire.

Though there was, in fact, nothing that could justly be considered as surprising in this event, yet I acknowledge fairly that it afforded me a degree of childish pleasure, which, were I ambitious of the reputation of a *grave philosopher*, I ought most certainly rather to hide than to discover.

The quantity of Heat excited and accumulated in this experiment was very considerable; for, not only the water in the box, but also the box itself (which weighed $15\frac{1}{4}$ lb.), and the hollow metallic cylinder, and that part of the iron bar which, being situated within the cavity of the box, was immersed in the water, were heated 150 degrees of Fahrenheit's scale; *viz.* from 60° (which was the temperature of the water and of the machinery at the beginning of the experiment) to 210°, the Heat of boiling water at Munich.

The total quantity of Heat generated may be estimated with some considerable degree of precision as follows:—

	Quantity of ice-cold water which, with the given quantity of Heat, might have been heated 180 degrees, or made to boil. In avoirdupois weight.
Of the Heat excited there appears to have been actually accumulated—	
In the water contained in the wooden box, $18\frac{3}{4}$ lb., avoirdupois, heated 150 degrees, namely, from 60° to 210°F. .	lb. 15.2
In 113.13 lb. of gun-metal (the hollow cylinder), heated 150 degrees; and, as the capacity for Heat of this metal is to that of water as 0.1100 to 1.0000, this quantity of Heat would have heated $12\frac{1}{2}$ lb. of water the same number of degrees .	10.37
In 36.75 cubic inches of iron (being that part of the iron bar to which the borer was fixed which entered the box), heated 150 degrees; which may be reckoned equal in capacity for Heat to 1.21 lb. of water	1.01
N.B. No estimate is here made of the Heat accumulated in the wooden box, nor of that dispersed during the experiment.	
Total quantity of ice-cold water which, with the Heat actually generated by friction, and accumulated in 2 hours and 30 minutes, might have been heated 180 degrees, or made to boil .	26.58

From the knowledge of the *quantity* of Heat actually produced in the foregoing experiment, and of the *time* in which it was generated, we are enabled to ascertain *the velocity of its production*, and to determine how large a fire must have been, or how much fuel must have been consumed, in order that, in burning equably, it should have produced by combustion the same quantity of Heat in the same time.

In one of Dr. Crawford's experiments (see his Treatise on Heat, p. 321), 37 lb. 7 oz., Troy, = 181,920 grains of water, were heated $2\frac{1}{10}$ degrees of Fahrenheit's thermometer with the Heat generated in the combustion of 26 grains of wax. This gives 382,032 grains of water heated 1 degree with 26 grains of wax, or $14{,}693\frac{14}{26}$ grains of water heated 1 degree, or $\frac{14693}{180} = 81.631$ grains heated 180 degrees, with the Heat generated in the combustion of 1 grain of wax.

The quantity of ice-cold water which might have been heated 180 degrees with the Heat generated by friction in the before-mentioned experiment was found to be 26.58 lb., avoirdupois, = 188,060 grains; and, as 81,631 grains of ice-cold water require the Heat generated in the combustion of 1 grain of wax to heat it 180 degrees, the former quantity of ice-cold water, namely 188,060 grains, would require the combustion of no less than 2303.8 grains (= $4\frac{8}{10}$ oz., Troy) of wax to heat it 180 degrees.

As the experiment (No. 3) in which the given quantity of Heat was generated by friction lasted 2 hours and 30 minutes, = 150 minutes, it is necessary, for the purpose of ascertaining how many wax candles of any given size must burn together, in order that in the combustion of them the given quantity of Heat may be generated in the given time, and consequently *with the same celerity* as that with which the Heat was generated by friction in the experiment, that the size of the candles should be determined, and the quantity of wax consumed in a given time by each candle in burning equably should be known.

Now I found, by an experiment made on purpose to finish these computations, that when a good wax candle, of a moderate size, $\frac{3}{4}$ of an inch in diameter, burns with a clear flame, just 49 grains of wax are consumed in 30 minutes. Hence it appears that 245 grains of wax would be consumed by such a candle in 150 minutes;

and that, to burn the quantity of wax (=2303.8 grains) necessary to produce the quantity of Heat actually obtained by friction in the experiment in question, and in the given time (150 minutes), *nine candles*, burning at once, would not be sufficient; for 9 multiplied into 245 (the number of grains consumed by each candle in 150 minutes) amounts to no more than 2205 grains; whereas the quantity of wax necessary to be burnt, in order to produce the given quantity of Heat, was found to be 2303.8 grains.

From the result of these computations it appears, that the quantity of Heat produced equably, or in a continual stream (if I may use that expression), by the friction of the blunt steel borer against the bottom of the hollow metallic cylinder, in the experiment under consideration, was *greater* than that produced equably in the combustion of *nine wax candles*, each $\frac{3}{4}$ of an inch in diameter, all burning together, or at the same time, with clear bright flames.

As the machinery used in this experiment could easily be carried round by the force of one horse (though, to render the work lighter, two horses were actually employed in doing it), these computations shew further how large a quantity of Heat might be produced, by proper mechanical contrivance, merely by the strength of a horse, without either fire, light, combustion, or chemical decomposition; and, in a case of necessity, the Heat thus produced might be used in cooking victuals.

But no circumstances can be imagined in which this method of procuring Heat would not be disadvantageous; for more Heat might be obtained by using the fodder necessary for the support of a horse as fuel.

As soon as the last-mentioned experiment (No. 3) was finished, the water in the wooden box was let off, and the box removed; and the borer being taken out of the cylinder, the scaly metallic powder which had been produced by the friction of the borer against the bottom of the cylinder was collected, and, being carefully weighed, was found to weigh 4145 grains, or about $8\frac{2}{3}$ oz., Troy.

As this quantity was produced in $2\frac{1}{2}$ hours, this gives 824 grains for the quantity produced *in half an hour*.

In the first experiment, which lasted only *half an hour*, the quantity produced was 837 grains.

In the experiment No. 1, the quantity of Heat generated in *half an hour* was found to be equal to that which would be required to heat 5 lb., avoirdupois, of ice-cold water 180 degrees, or cause it to boil.

According to the result of the experiment No. 3, the Heat generated in *half an hour* would have caused 5.31 lb. of ice-cold water to boil. But, in this last-mentioned experiment, the Heat generated being more effectually confined, less of it was lost; which accounts for the difference of the results of the two experiments.

It remains for me to give an account of one experiment more, which was made with this apparatus. I found, by the experiment No. 1, how much Heat was generated when the air had free access to the metallic surfaces which were rubbed together. By the experiment No. 2, I found that the quantity of Heat generated was not sensibly diminished when the free access of the air was prevented; and by the result of No. 3, it appeared that the generation of the Heat was not prevented or retarded by keeping the apparatus immersed in water. But as, in this last-mentioned experiment, the water, though it surrounded the hollow metallic cylinder on every side, externally, was not suffered to enter the cavity of its bore (being prevented by the piston), and consequently did not come into contact with the metallic surfaces where the Heat was generated; to see what effects would be produced by giving the water free access to these surfaces, I now made the

Experiment No. 4

The piston which closed the end of the bore of the cylinder being removed, the blunt borer and the cylinder were once more put together; and the box being fixed in its place, and filled with water, the machinery was again put in motion.

There was nothing in the result of this experiment that renders it necessary for me to be very particular in my account of it. Heat was generated as in the former experiments, and, to all appearance, quite as rapidly; and I have no doubt but the water in the box would have been brought to boil, had the experiment been continued as long as the last. The only circumstance that surprised me was, to find how little difference was occasioned in the noise made by the borer in rubbing against the bottom of the bore of the cylinder, by filling the bore with water. This noise, which was very grating to the ear, and sometimes almost insupportable, was, as nearly as I could judge of it, quite as loud and as disagreeable when the surfaces rubbed together were wet with water as when they were in contact with air.

By meditating on the results of all these experiments, we are naturally brought to that great question which has so often been the subject of speculation among philosophers; namely,—

What is Heat? Is there any such thing as an *igneous fluid*? Is there anything that can with propriety be called *caloric*?

We have seen that a very considerable quantity of Heat may be excited in the friction of two metallic surfaces, and given off in a constant stream or flux *in all directions* without interruption or intermission, and without any signs of diminution or exhaustion.

From whence came the Heat which was continually given off in this manner in the foregoing experiments? Was it furnished by the small particles of metal, detached from the larger solid masses, on their being rubbed together? This, as we have already seen, could not possibly have been the case.

Was it furnished by the air? This could not have been the case; for, in three of the experiments, the machinery being kept immersed in water, the access of the air of the atmosphere was completely prevented.

Was it furnished by the water which surrounded the machinery? That this could not have been the case is evident: *first*, because this water was continually *receiving Heat* from the machinery, and could not at the same time be *giving to*, and *receiving Heat from*, the same body; and, *secondly*, because there was no chemical decomposition of any part of this water. Had any such decomposition taken place (which, indeed, could not reasonably have been expected), one of its component elastic fluids (most probably inflammable air) must at the same time have been set at liberty, and, in making its escape into the atmosphere, would have been detected; but though I frequently examined the water to see if any air-bubbles rose up through it, and had even made preparations for catching them, in order to examine them, if any should appear, I could perceive none; nor was there any sign of decomposition of any kind whatever, or other chemical process, going on in the water.

Is it possible that the Heat could have been supplied by means of the iron bar to the end of which the blunt steel borer was fixed? or by the small neck of gun-metal by which the hollow cylinder was united to the cannon? These suppositions appear more improbable even than either of those before mentioned; for Heat was continually going off, or *out of the machinery*, by both these passages, during the whole time the experiment lasted.

And, in reasoning on this subject, we must not forget to consider that most remarkable circumstance, that the source of the Heat generated by friction, in these experiments, appeared evidently to be *inexhaustible.*

It is hardly necessary to add, that anything which any *insulated* body, or system of bodies, can continue to furnish *without limitation*, cannot possibly be *a material substance*; and it appears to me to be extremely difficult, if not quite impossible, to form any distinct idea of anything capable of being excited and communicated in the manner the Heat was excited and communicated in these experiments, except it be MOTION.

I am very far from pretending to know how, or by what means or mechanical contrivance, that particular kind of motion in bodies which has been supposed to constitute Heat is excited, continued, and propagated; and I shall not presume to trouble the Society with mere conjectures, particularly on a subject which, during so many thousand years, the most enlightened philosophers have endeavoured, but in vain, to comprehend.

But, although the mechanism of Heat should, in fact, be one of those mysteries of nature which are beyond the reach of human intelligence, this ought by no means to discourage us or even lessen our ardour, in our attempts to investigate the laws of its operations. How far can we advance in any of the paths which science has opened to us before we find ourselves enveloped in those thick mists which on every side bound the horizon of the human intellect? But how ample and how interesting is the field that is given us to explore!

Nobody, surely, in his sober senses, has even pretended to understand the mechanism of gravitation; and yet what sublime discoveries was our immortal Newton enabled to make, merely by the investigation of the laws of its action!

The effects produced in the world by the agency of Heat are probably *just as extensive*, and quite as important, as those which are owing to the tendency of the particles of matter towards each other; and there is no doubt but its operations are, in all cases, determined by laws equally immutable.

Before I finish this Essay, I would beg leave to observe, that although, in treating the subject I have endeavoured to investigate, I have made no mention of the names of those who have gone over the same ground before me, nor of the success of their labours, this omission has not been owing to any want of respect for my predecessors, but was merely to avoid prolixity, and to be more at liberty to pursue, without interruption, the natural train of my own ideas.

Description of the Figures

[Plate 5, Figs. 1–3; Plate 6, Figs. 4–8]

Fig. 1 shews the cannon used in the foregoing experiments in the state it was in when it came from the foundry.

Fig. 2 shews the machinery used in the experiments No. 1 and No. 2. The cannon is seen fixed in the machine used for boring cannon. W is a strong iron bar (which, to save room in the drawing, is represented as broken off), which bar, being united with machinery (not expressed in the figure) that is carried round by horses, causes the cannon to turn round its axis.

m is a strong iron bar, to the end of which the blunt borer is fixed; which, by being forced against the bottom of the bore of the short hollow cylinder that remains connected by a small cylindrical neck to the end of the cannon, is used in generating Heat by friction.

Fig. 3 shews, on an enlarged scale, the same hollow cylinder that is represented on a smaller scale in the foregoing figure. It is here seen connected with the wooden box (*g*, *h*, *i*, *k*) used in the experiments No. 3 and No. 4, when this hollow cylinder was immersed in water.

p, which is marked by dotted lines, is the piston which closed the end of the bore of the cylinder.

n is the blunt borer seen sidewise.

d, *e*, is the small hole by which the thermometer was introduced that was used for ascertaining the Heat of the cylinder. To save room in the drawing, the cannon is represented broken off near its muzzle; and the iron bar to which the blunt borer is fixed is represented broken off at *m*.

Fig. 4 is a perspective view of the wooden box, a section of which is seen in the foregoing figure. (See *g*, *h*, *i*, *k*, Fig. 3.)

Figs. 5 and 6 represent the blunt borer *n*, joined to the iron bar *m*, to which it was fastened.

Fig. 7 and 8 represent the same borer, with its iron bar, together with the piston which, in the experiments No. 2 and No. 3, was used to close the mouth of the hollow cylinder.

Count Rumford realized that his experiments would not go unchallenged. In fact he raised a number of questions in the paper just given in order to answer them himself. A number of writers in the field disagreed with Rumford's conclusions, and to give the flavor of their objections let us quote from a paper by Emmett published in the *Annals of Philosophy* **16**, 137 (1820).

First, in answer to Rumford's question "Is the heat furnished by the metallic chips which are separated by the borer from the solid mass of metal?" Emmett comments:

> "In the commencement of this reasoning, an assumption is made, which is particularly unfortunate: namely, that if heat being an elastic fluid be evolved by the compression of solid matter, the capacity of that solid for heat must be diminished in proportion to the quantity which has been separated. The whole quantity of heat contained in the solid is doubtless diminished, but why is the capacity to be changed? . . . That the quantity of heat evolved in this experiment was great cannot be disputed, yet it was

by no means sufficient to warrant the conclusions that have been drawn. . . . In these experiments, a very large mass of metal was submitted to an excessive pressure and of the mass, fresh strata was continually exposed to the compression by the wearing off of the brass: hence a definite quantity of heat was separated from each stratum in succession. Now if we admit the existence of caloric in a state of great density in the metals, this cause would be quite adequate to the production of the observed effect. The greatest error appears to be the assumption that the source of the heat thus generated is inexhaustible; the quantity that can be thus excited is finite"

but will not cease, according to this pictures, until all the brass is worn away.

Let us return to more of Rumford's questions. "From whence came the Heat which was continually given off in this manner in the foregoing experiments? Was it furnished by the air, was it furnished by the water which surrounded the machinery? Is it possible that the heat could have been supplied by means of the iron bar to the end of which the blunt steel borer was fixed? Or the small neck of gun-metal by which the hollow cylinder was united to the cannon? These suppositions appear most improbable." The answer to these questions seemed to lie at the very foundation of the caloric theory, namely that caloric pervaded all matter, and therefore could be furnished by all those sources which the Count appeared to discredit. His whole apparatus was bathed in an atmosphere of caloric. A physical picture of the manner in which the caloric was resupplied to the brass cup during the experiment was described as follows by a correspondent to the *Philosophical Magazine* for 1816.†

"If heat be a material fluid, the effect of force on a body containing it would be similar to the force on a body containing any other fluid diffused through its pores in a similar manner. Water being a fluid which in many instances produces effects similar to those produced by heat, it appears best adapted to illustrate the generation of heat by friction.

"I procured a piece of light and porous wood . . . and having immersed it in water until it was saturated, I fixed it firmly over a vessel filled with water, the lower end being . . . below the surface of the water, and then moved a piece of hard wood backwards and forwards on the upper end, with a considerable degree of pressure. I thus found that water could be raised through the pores of wood by friction. The process is easily understood: the piece of hard wood, as it is moved along, presses the water out of the pores, and closes them, driving out the water which is pressed out before it; but when the hard wood has pressed over these pores, the water from below rushes into them to restore the equilibrium.

"The action of the blunt borer in Count Rumford's experiments appears to have produced a similar kind of effect, the heat having been forced out of the pores of the metal by the borer, its place would be supplied by heat from the adjacent parts. Gun-metal being a good conductor, the neck which connected the cylinder with the cannon would be capable of giving passage to all the heat which was accumulated from the cannon, and the other conductors with which it was connected."

† *Phil. Mag.* **48,** 29 (1816).

CHAPTER 5

THERMAL EXPANSION

THE caloric theory supplied an obvious answer to the problem of thermal expansion and contraction. Heating a body consisted merely of adding the fluid caloric to the body. Since the fluid occupied space, one would expect an expansion. On cooling, the caloric fluid was removed and hence the body contracted. The difference between solids, liquids, and gases was considered to lie essentially in the degree of gravitational attraction between the atoms of the substance. With small amounts of heat, the caloric repulsion was not strong and the atoms were tightly bound by strong gravitational attraction. As the temperature of a body was increased, the attraction became less as the thermal repulsion became greater. In a liquid, the caloric content was sufficiently high so that the atoms were not held in a rigid position by the mutual gravitational attraction. In a gas, the gravitational attraction was considered inoperative. This theory, therefore, led one to expect the observed effect that the expansion of a gas was greater than in the case of a liquid, which in turn was greater than that of a solid.

According to the caloric theory, when a substance was cooled and the caloric fluid extracted from it, it contracted and hence became denser. It is, therefore, not surprising to find Rumford studying carefully the case of water near its freezing point, where its change in density is anomalous. In his "Account of Some New Experiments on the Temperature of Water at its Maximum Density" he showed that water at 41°F is denser than water at 32°F by demonstrating that water at this higher temperature sinks in water at the lower temperature.

In other parts of this paper, and also in a paper entitled "An Account of a Curious Phenomenon Observed on the Glaciers of Chamouny",† Count Rumford arrived at the temperature of 41°F as that for the maximum density of water. His primary purpose in carrying out these experiments, however, is expressed in the conclusion of the paper on glaciers:

> These experiments ought not to be regarded as suitable for determining with great exactness the temperature at which the density of water is at a maximum, but rather as proving that this temperature is really several degrees of the thermometric scale above that of melting ice.

Count Rumford offered no explanation of this phenomenon using the energy theory of heat. However, he felt that the fact that the removal of the

† *Philosophical Transactions* **94,** 23 (1804).

fluid caloric did not cause the water continuously to contract was a serious objection to the material theory.

An account of some new Experiments on the Temperature of Water at its Maximum Density

Read to the National Institute of France
July 15, 1805

Memoires de l'Institut VII, 78–97 (1806)
Nicholson's Journal XI, 225–235 (1805)
Tilloch's Philosophical Magazine XXVI, 273 (1805)
Gilbert's Annalen der Physik XX, 369–383 (1805)
Bibliotheque Britannique XXXIV, 113–120 (1806)

In my seventh Essay on the Propagation of Heat in Fluids, and in a paper published in the Philosophical Transactions for the year 1804, in which I have given an account of a curious phenomenon frequently observed on the glaciers of Chamouny, I have ascribed the melting of the ice below the surface of the ice-cold water to currents of water slightly warmer, and consequently slightly heavier, which descend from the surface to the bottom of the ice-cold water; but the principal fact on which this supposition is founded having been called in question by various persons, I have endeavoured to establish it by new and decisive experiments.

If it is true that the temperature of water at its maximum density is considerably higher than the freezing-point of that liquid (as was announced many years ago by M. de Luc), and that the communication of heat in liquids is brought about by a movement of circulation caused by a change of density in the particles of the fluid resulting from a change of temperature, the explanation that I have given of the phenomenon of the melting of ice covered with a layer of ice-cold water by heat applied to the surface of the water, would seem natural and admissible; but if the density of water is greater at the temperature of melting ice than at any other more elevated temperature, as some philosophers assert, it is evident that the vertical descending currents of warm water which I have described cannot exist, and my explanation must be rejected.

This inquiry interested me all the more, because the fact in question had served as the foundation of the theory which I gave in my seventh Essay on the periodical winds of the polar regions, and as the basis of my conjectures on the existence of currents of cold water in the depths of the sea coming from the polar regions to the equator, and on the cause of the great difference which is

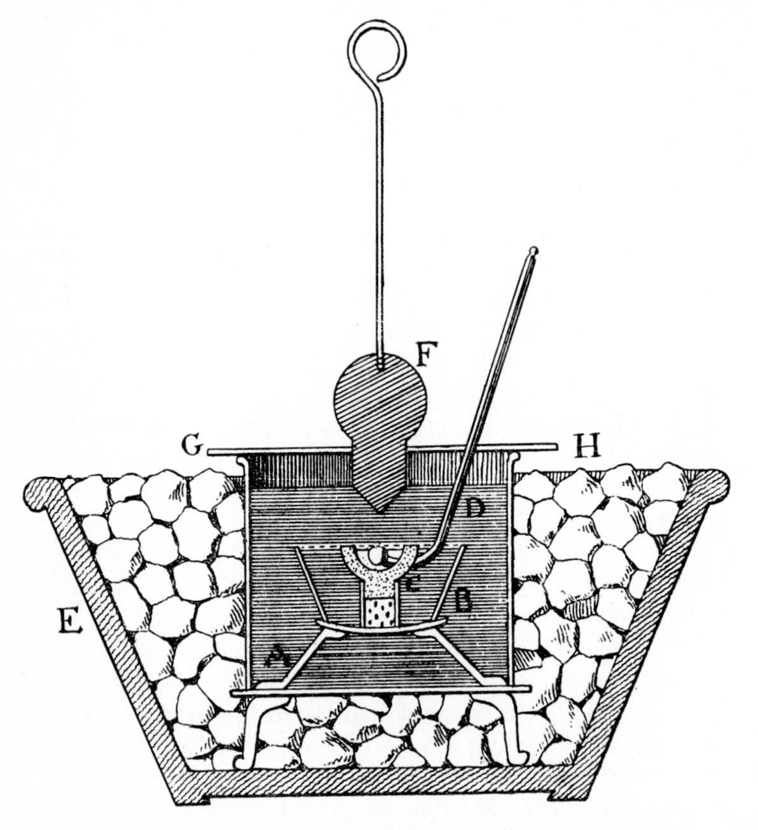

PLATE 7. Experimental arrangement for measuring the temperature of water at its maximum density.

found in the temperature of different countries situated in the same latitude and at the same height above the level of the sea.

After meditating on the means which I should employ to establish this important fact beyond doubt, I thought of the experiment which I am about to describe, and which is all the more interesting, since it not only demonstrates the existence in a mass of water which is warmed or cooled, of the currents assumed by my theory, but proves at the same time that the temperature at which the density of water is at a maximum is actually some degrees above that of melting ice.

Having provided a cylindrical vessel (A, Plate 7), open above, made of thin sheet brass, $5\frac{1}{2}$ inches in diameter and 4 inches deep, supported on three strong legs $1\frac{1}{4}$ inches high, I placed in it a thin brass cup (B) 2 inches in diameter at its bottom (which is a little convex downwards), $2\frac{8}{10}$ inches wide at its brim, and $1\frac{3}{10}$ inches deep; this cup stands on three spreading legs made of strong brass wire, and of such form and length that when the cup is introduced into the cylindrical vessel, it remains firmly fixed in the axis of it, and in such a situation that the bottom of the cup is elevated just $1\frac{1}{4}$ inches above the bottom of the cylindrical vessel.

In the middle of this cup there stands a vertical tube of thin sheet brass $\frac{1}{2}$ of an inch in diameter and $\frac{6}{10}$ of an inch in length, open above, which serves as a support for another smaller cup (C), which is made of cork, the brim of which is on the same horizontal level with the brim of the larger brass cup in which it is placed.

This cork cup, which is spherical, being something less than half of an hollow sphere, is 1 inch in diameter at its brim, measured within, $\frac{4}{10}$ of an inch deep, and $\frac{1}{4}$ of an inch in thickness. It is firmly attached to the vertical tube on which it stands, by means of a cylindrical foot $\frac{1}{2}$ of an inch in diameter and $\frac{1}{4}$ of an inch high, which enters with friction into the opening of the vertical tube.

On one side of this cork cup there is a small opening, which receives and in which is confined the lower extremity of the tube of a small mercurial thermometer (D). The bulb of this thermometer, which is spherical, is $\frac{3}{10}$ of an inch in diameter, and it is so fixed in the middle of the cup, that its centre is $\frac{1}{4}$ of an inch above the bottom of the cup; consequently it does not touch the cup anywhere, nor does any part of it project above the level of its brim.

The tube of this thermometer, which is 6 inches in length, has an elbow near its lower end at the distance of 1 inch from its bulb, which elbow forms an angle of about 110 degrees, and the thermometer is so fixed in the cork cup, that the short branch of its tube, namely, that to the end of which the bulb is attached, lies in an horizontal position, while the longer branch (to which a scale made of ivory and graduated according to Fahrenheit is affixed) projects obliquely upwards and outwards in such a manner that the freezing-point of the scale lies just above the level of the top of the cylindrical vessel in which the cups are placed.

The cork cup, which was turned in the lathe, is neatly formed, and in order to close the pores of the cork, it was covered within and without with a thin coating of melted wax, which was polished after the wax was cold.

The thermometer was fixed to the cork cup by means of wax, and in doing this care was taken to preserve the regular form of the cup, both within and without.

The vertical brass tube which supports this cup in the axis of the brass cup is pierced with several small holes, in order to allow the water employed in the experiments to pass freely into and through it.

Having attached about 6 ounces of lead to each of the legs of the brass cup, in order to render it the more steady in its place, it was now introduced with its contents into the cylindrical vessel, and the vessel was placed in an earthen basin (E), and surrounded on all sides with pounded ice. This basin was 11 inches in diameter at its brim, 7 inches in diameter at the bottom, and 5 inches deep, and was placed on a firm table in a quiet room.

Several cakes of ice were then placed under the bottom of the brass cup, and the cup was surrounded on all sides by a circular row of other long pieces of ice fixed in a vertical position between the outer walls of the cup and the walls of the cylindrical vessel.

These pieces were about 4 inches long, and extended from the bottom of the vessel to within a very short distance of the top. All these pieces of ice having been fixed firmly in their places by means of some little wooden wedges, ice-cold water was poured into the cylindrical vessel until the surface of this liquid was an inch above the upper edge of the cork cup.

In this state of things it is evident that the two cups were filled with and surrounded on all sides by water at the temperature of melting ice, and that this temperature was maintained constant by the pieces of ice with which the water was in contact.

After having left the apparatus in this situation for about an hour, in order to satisfy myself that the temperature of the cold water was constant and uniform throughout its entire mass, I made the following experiment.

Experiment No. 1.—A solid ball of tin (F) having been provided, 2 inches in diameter, with a cylindrical projection on the lower side of it, 1 inch in diameter and ½ of an inch long ending in a conical point which projected (downwards) ½ of an inch farther, and having on the other side a strong iron wire 6 inches long, which served as a handle,—this ball, after having been immersed for half an hour in a considerable quantity of water at the temperature of 42°F., was withdrawn from the water, wiped dry with a handkerchief of the same temperature, placed without loss of time above the cylindrical vessel, and fixed in such a position that the entire conical point of the tin ball (½ of an inch in length) was submerged in the cold water contained in the vessel.

To fix and keep the metallic ball in its place, I used a strong slip of tin (GH), 6 inches long and 2½ inches wide, with a circular hole in the middle of it 1 inch in diameter. This slip of tin being laid horizontally on the top or brim of the cylindrical vessel in such a manner that the centre of the circular hole coincided with the axis of the cylindrical vessel, when the short cylindrical projection belonging to the ball was introduced into that hole, the ball was firmly supported in its proper place.

The ball was placed in such a position that the end of the conical projection was immediately over the cork cup, at the distance of

$1\frac{1}{2}$ inches above the level of its brim and consequently $\frac{1}{2}$ of an inch above the upper part of the bulb of the small thermometer which lay in this cup.

The quantity of cold water in the cylindrical vessel had been so regulated beforehand that when the conical point was entirely submerged, the surface of the water was on a level with the base of this inverted cone, so that the whole of the cylindrical part of the projection was out of the water.

I knew that the particles of ice-cold water which were thus brought into contact with the conical point could not fail to acquire some small degree of heat from that relatively warm metal, and I concluded that if the particles of water so warmed should in fact become *heavier* than they were before, in consequence of this small increase of temperature, they must necessarily *descend* in the surrounding lighter ice-cold liquid, and as the heated metallic point was placed directly over the cork cup, and fixed immovably in that situation, I foresaw that the descending current of warm water must necessarily fall into that cup and at length fill it, and that the presence of this warm water in the cup would be announced by the rising of the thermometer.

The result of this very interesting experiment was just what I expected: the conical metallic point had not been in contact with the ice-cold water more than 20 seconds when the mercury in the thermometer began to rise, and in 3 minutes it had risen three degrees and a half, namely, from 32° to $35\frac{1}{2}$°; when 5 minutes had elapsed it had risen to 36°.

Another small thermometer placed just below the surface of the ice-cold water, and only $\frac{2}{10}$ of an inch from the upper part of the conical point and on one side of it, did not appear to be sensibly affected by the vicinity of that warm body.

A third thermometer, the bulb of which was placed in the brass cup just on the outside of the cork cup and on a level with its brim, showed that the water which immediately surrounded the cork cup remained constantly at the temperature of freezing during the whole time that the experiment lasted.

As I well knew from the results of the experiments on the propagation of heat in a solid bar of metal,* that no one of the particles of cold water in contact with the surface of the conical projection, in the experiment which I have just described, could acquire by this momentary contact a temperature as high as that of the warm metal, I was by no means surprised to find that the thermometer belonging to the cork cup rose no higher than 36°.

In order to see if it could not be made to rise not only higher, but also more rapidly, by employing the metallic ball heated to such a temperature as it might be supposed would be sufficient to heat those particles of ice-cold water which should come into contact with its conical point, to the temperature at which the density of water is supposed to be a maximum, I made the following experiment.

Experiment No. 2.—Having removed the ball, I gently brushed away the warm water which in the last experiment had been lodged in the cavity of the cork cup, and which still remained there, as was evident from the indication of the thermometer belonging to the cup; I then placed several small cakes of ice in the cylindrical vessel, which ice, floating on the surface of the water in the vessel, prevented the water from receiving heat from the surrounding air, which at that time was at the temperature of 70°F. As the cork cup had been a little heated by the warm water in the foregoing experiment, time was now given it to cool.

As soon as the cup and the whole mass of the water in the cylindrical vessel appeared to have acquired the temperature of freezing, I carefully removed the cakes of ice which floated on the surface of the water, and introduced once more the projecting conical point belonging to the metallic ball into the ice-cold water in the vessel, placing it exactly in the same place which it had occupied in the foregoing experiment; but this ball, instead of being at the temperature of 42°F., as before, was now at the temperature of 60°F.

The results of this experiment were very striking, and, if I am

* An account of these experiments has been given in a memoir presented to the Mathematical and Physical Class of the National Institute of France, on the 7th of May, 1804.

not much mistaken, afford a direct, unexceptionable, and demonstrative proof, not only that the maximum of the density of water is in fact at a temperature which is several degrees above the point of freezing, but also that warm currents do actually set downwards in ice-cold water, whenever a certain small degree of heat is applied to the particles of that fluid which are at its surface, as I have already announced in my Essay on the Propagation of Heat in Fluids.

The conical metallic point had been in its place no more than 10 seconds when I distinctly saw that the mercury in the thermometer belonging to the cork cup was in motion, and, when 50 seconds had elapsed, it had risen four degrees, *viz*. from 32° to 36°.

When 2 minutes and 30 seconds had elapsed, reckoning from the moment when the metallic point was introduced into the cold water, the thermometer had risen to 39°, and at the end of 6 minutes to $39\frac{7}{8}$°, when it began to fall; but very slowly, however, for at the end of 8 minutes and 30 seconds it was at $39\frac{3}{4}$°.

A small mercurial thermometer, the bulb of which was placed on one side of the cork cup at the distance of about $\frac{2}{10}$ of an inch from it, showed no signs of being in the least affected by the vertical current of warm water which descended from the conical point into the cup in this experiment.

This experiment was repeated four times the same day (the 13th of June, 1805), and always with nearly the same results. The mean results of these four experiments were as indicated on the next page.

As I had found by some of my experiments made in the year 1797 (of which an account is given in my seventh Essay, Part I.) that water at the temperature of about 42°F., and consequently what we should call very cold, melted considerably more ice, when standing on it, than an equal quantity of boiling-hot water in the same situation, I was very curious to see whether the thermometer, the bulb of which lay in the cork cup, would not also be less heated by the ball when it should be applied *very hot* to the surface of the water, than when its temperature was much lower.

Time elapsed, reckoned from the beginning of the experiment			Temperature of the water in the cork cup, as shown by the thermometer
m.	s.		Degrees
0	0	. . . .	32
At 0	10	began to rise	32+
At 0	23	had risen to	33
0	28	,, ,, ,, ...	34
0	35	,, ,, ,, ...	35
0	48	,, ,, ,, ...	36
1	3	,, ,, ,, ...	37
1	35	,, ,, ,, ...	38
2	32	,, ,, ,, ...	39
3	41	,, ,, ,, ...	$39\frac{1}{2}$
4	48	,, ,, ,, ...	$39\frac{3}{4}$
6	5	,, ,, ,, ...	$39\frac{7}{8}$

Seeing that this research ought to throw great light on the mysterious operations of the distribution of heat in liquids, I hastened to make the following experiment.

Experiment No. 3.—The cylindrical vessel with its contents having been once more reduced to the uniform temperature of freezing water, the metallic ball was heated in boiling water, and being as expeditiously as possible taken out of that hot liquid, its projecting conical point was suddenly submerged in the ice-cold water, as in the former experiments.

The result of this experiment was very interesting. It was not till 50 seconds had elapsed that the thermometer began to show any signs of rising, and at the end of 1 minute and 7 seconds it had risen only 2 degrees.

In the foregoing experiment, when the metallic ball was so much colder, the thermometer began to rise in 10 seconds, and at the end of 1 minute and 3 seconds it had risen 5 degrees.

This difference is very remarkable, and if it does not prove the existence and great efficacy of currents in conveying heat in fluids, I must confess that I do not see how the existence of any invisible mechanical operation, the progress of which does not immediately fall under the cognizance of our senses, can ever be demonstrated.

As the experiment made with the ball heated in boiling water appeared to me to be very interesting, I repeated it twice, and its results were always nearly the same. The mean results of these three experiments were as follows:—

Time elapsed, reckoned from the beginning of the experiment			Temperature of the water in the cork cup, as shown by the thermometer
m.	s.		Degrees
0	0	. . . .	32
At 0	50	the thermometer began to rise .	32+
At 1	2	had risen to	33
1	7	” ” ”	34
1	18	” ” ”	35
2	2	” ” ”	36
3	2	” ” ”	$36\frac{1}{2}$
4	17	” ” ”	37
6	12	” ” ”	38
7	17	” ” ”	$38\frac{1}{8}$
9	0	” ” ”	$38\frac{1}{4}$
12	0	” ” ”	$38\frac{1}{4}$
14	0	” ” ”	$38\frac{1}{4}$

By comparing the mean results of these experiments with the mean results of those in which the ball was at the temperature of 60° or less, we may see how much more rapid the communication of heat in the cold water from above downwards was when the metallic ball was *relatively cold* than when it was much warmer; but we must not consider of too much importance the determination of the relative rapidity thus made, because it is more than probable that it was not till after the conical metallic point had been considerably cooled by contact with the cold water that the vertical descending currents could exist by which the thermometer was at length heated. At the beginning of the experiment made with the tin ball warmed in boiling water, the particles of water which were in immediate contact with the conical point while it was still very warm, were heated to a temperature higher than that at which the density of water is at a maximum, and the density of these particles being diminished by this high degree of heat, the vertical currents in the cold water were at the beginning ascending currents, as I satisfied myself by means of a small thermometer placed by the side of the conical point at a distance of $\frac{2}{10}$ of an inch from its base, and immediately below the surface of the cold water: this thermometer began to rise very rapidly as soon as the warm metallic point was plunged into the cold water.

Another small thermometer, the bulb of which was situated at about the same distance from the axis of the conical projection, but $\frac{1}{2}$ of an inch below the surface of the cold water, preserved throughout the entire experiment the appearance of perfect rest.

The results of this last experiment are all the more interesting because they afford a demonstrative proof that it was neither by a direct communication of heat in the water, which was at rest, from molecule to molecule, *de proche en proche*, nor by calorific radiations passing through the water, that heat was communicated from the metallic point to the bulb of the thermometer, but actually by a descending current of warm water; for it is perfectly evident that if this heat had been communicated either by a direct transfer in the water from molecule to molecule or by calorific radiations passing from the surface of the metal through the water, which remained at rest, this communication would naturally have been the most rapid when the metallic point was the warmest. What did take place was exactly contrary to this, as we have just seen. Moreover, the small thermometer, which was placed close to the metallic body on one side, and which in this experiment was in no degree affected by the heat of this body, would not have failed to acquire as much heat at least as that placed in the cork cup, which was situated below the metallic body and at a greater distance from it.

The considerable amount of time which elapsed in the experiments performed with the tin ball heated in boiling water before the thermometer in the cork cup began to be so sensibly affected, and the rapidity with which it was then warmed through several degrees as soon as it began to rise, indicated a fact which it is important to notice. In order to throw light upon this fact, we must consider carefully the operation of the heating of cold water by the warm metallic surface with which it was in contact, and examine it in its progress and in all its details.

Let us begin by supposing that the conical point of the ball, at the temperature of boiling water, has just been submerged vertically up to the level of its base in a mass of undisturbed water at the temperature of melting ice. As the particles of water, which in this case are in contact with the warm metallic surface, cannot pass, all of a sudden, from the temperature of melting ice to that of boiling water without passing through all the intermediate degrees, and since these particles at the temperature of melting ice cannot become warmer without becoming more dense, it is evident that they must have a tendency to descend, and consequently to leave the surface of the metal, as soon as they begin to acquire heat; but experiment showed that, instead of descending, they were actually pushed upwards: this proves that they were heated so rapidly that, before they had time to leave the surface of the metal and to escape from its calorific influence, they had acquired a temperature so elevated that their density, after having passed rapidly the point of its maximum, became even less than it was at the temperature of melting ice. But after some moments, the metallic body having cooled somewhat, and the communication of heat to the particles of water taking place more slowly, these particles, having become more dense on account of a slight increase of temperature had time to escape before becoming warmer, and at that time the descending current suddenly began.

This fact interests me the more, as it may serve in some sort to explain a phenomenon which I observed in an experiment made eight years ago, an account of which I gave in my Essay on the Propagation of Heat in Fluids.

The phenomenon to which I have alluded was this: Having poured some mercury into a small cylindrical glass vessel 2 inches in diameter and $3\frac{1}{2}$ inches deep, until this fluid filled the vessel to the height of an inch, I poured on to the mercury twice as much water (that is, 2 inches), and plunging the vessel up to the level of the upper surface of the mercury into a freezing mixture of pounded ice and sea-salt, the temperature of the air being 60°F., I allowed the whole to cool quietly, in order to see in what part of the water the ice would first appear. It was at the bottom of the water, where this liquid was in contact with the mercury, that the ice formed.

The layer of water which rested immediately on the surface of the mercury having been cooled to about the temperature of 41°F., where the density of water is at its maximum, the particles of this water, which were then in immediate contact with the mercury, losing still more of their heat, became of necessity less dense, and had consequently a tendency to leave the bottom of the water and to ascend upwards; but the rapidity with which they were cooled by the mercury was so great that they were frozen before they could escape from the cooling influence of this cold body.

After all that I have said about the warm and cold currents which take place in a liquid which is warmed or cooled, it might perhaps be thought that I regard these currents as composed of single particles of the liquid, which, having been in immediate contact with the body which gives or which receives the heat, are all of the same temperature. I am all the farther from holding this opinion, since I know from the results of several experiments made expressly for elucidating this point, (and which I shall have the honor of presenting to the Class on another occasion,) that a liquid current cannot pass through another liquid mass which is at rest, and which is of the same kind and of about the same specific gravity, without producing a perceptible mixture of the two liquids; much less, therefore, can a small current of warm water pass without mixing through a mass of cold water; and the farther it advances the more it will be mixed, and the more, in consequence, will its temperature be found to be lowered.

For example, in the experiments of which I have just given an account, the cork cup, which received the current of warm water descending from the metallic point of the tin ball, was only $\frac{1}{2}$ of an inch below the extremity of this point; if this distance had been greater, the thermometer in the cup would certainly have risen to a less height: for this reason these experiments ought not to be regarded as suitable for determining with great exactness the

D

temperature at which the density of water is at a maximum, but rather as proving that this temperature is really several degrees of the thermometric scale above that of melting ice; and this is all that I am particularly interested in showing at the present time.

Judging from the constant temperature which is found at all seasons at the bottom of deep lakes and from the results of several direct experiments, we may conclude that water is at its *maximum* density when it is at the temperature of about 41° of Fahrenheit's scale, which corresponds to 4° on that of Reaumur, and to 5° of the Centigrade scale.

Some of the most eminent scientists of his day took issue with the results of these experiments of the Count. For example, John Dalton published a paper in 1805 under the title "Some Remarks Against the Experiments of Count R. on the Temperature at which Water is Densest".

Dalton writes in part:

> "However keen and worthy of attention this experiment of Count R. may be, yet I do not regard it as conclusive as Count R. seems to do. . . .
> "I am still of the opinion that water is most dense at the freezing point, 30°F., and I intend here to show how I explain the results of Count R. according to my hypothesis. Water expands by heat from any determined temperature, whatever it may be, nearly proportional to the square of the temperature. . . . Consequently the 'rising power' that the water receives from the heat, at first very small, increases nevertheless shortly before boiling to a considerable amount. The cohesion of the water particles is a constant force. Consequently, somewhere between these two forces, an equilibrium point will appear. That is, a point where the rising force through increasing temperature will be just sufficient to overcome the 'stickiness' of the water and, in this case, no inner motion can result . . . I presume that water, at 40°F, is about $\frac{1}{10000}$ lighter than water at 30°F, but that the resulting rising force is just equal to the cohesion of the water, wherefore at this temperature, no upward streaming results, and, in this case, the heat spreads out through the water just as through a solid body. . . .
>
> "Therefore, the experiment of Count R. can be explained by the fact that the thermometer in the experiment was heated by the peculiar conductivity characteristic of the water just as though the medium had been metal or some other solid body."

CHAPTER 6

THE WEIGHT OF HEAT

ONE of the characteristics of material substances is that it demonstrates its existence by having weight. On top of that, the explanation of many caloric phenomena was based on the strong attraction between the caloric fluid and matter. Because of this attraction, one would expect that the force between the caloric in a body and the earth, namely, the weight of caloric, should be a measurable quantity. One of Count Rumford's major investigations was an attempt to measure this weight of heat both in its "sensible" and "latent" form. This experiment is one of the cleverest ones which the Count conceived of, relying as it did both on the difference of specific heats in bodies as liquids were heated, and also on the latent heat of fusion when water froze. Many other attempts to measure the weight of heat had been undertaken with very conflicting results. Most of these other results suffered either from the disturbing effects of convection currents when bodies were heated or cooled on the arms of a balance, or they did not properly take into account the change of buoyancy due to Archimedes' principle with the expansion and contraction of the bodies themselves. In avoiding these difficulties by carrying out his experiment essentially at constant temperature and by utilizing the large heats of fusion of liquids, Count Rumford succeeded in demonstrating that within the limits of error of his balance, which was as good as could be obtained at the time, no change in weight was observed for very large changes in heat content.

An Inquiry concerning the Weight ascribed to Heat

Read before the Royal Society, May 2, 1799

Philosophical Transactions LXXXIX, 179–194 (1799)
Philosophical Magazine V, 162–174 (1799)
Rumford's Philosophical Papers, Vol. I, 366–383 (1802)
Bibliotheque Britannique XIII, 217–238 (1799)
Nicholson's quarto Journal III, 381–390 (1799)
Scherer's Journal der Chemie V, 53–70 (1800)

The various experiments which have hitherto been made with a view to determine the question, so long agitated, relative to the weight which has been supposed to be gained, or to be lost, by bodies upon their being heated, are of a nature so very delicate, and are liable to so many errors, not only on account of the imperfections of the instruments made use of, but also of those, much more difficult to appreciate, arising from the vertical currents in the atmosphere, caused by the hot or the cold body which is placed in the balance, that it is not at all surprising that opinions have been so much divided, relative to a fact so very difficult to ascertain.

It is a considerable time since I first began to meditate upon this subject, and I have made many experiments with a view to its investigation; and in these experiments I have taken all those precautions to avoid errors which a knowledge of the various sources of them, and an earnest desire to determine a fact which I conceived to be of importance to be known, could inspire; but though all my researches tended to convince me more and more that *a body acquires no additional weight upon being heated*, or, rather, that heat has no effect whatever upon the weights of bodies, I have been so sensible of the delicacy of the inquiry, that I was for a long time afraid to form a decided opinion upon the subject.

Being much struck with the experiments recorded in the Transactions of the Royal Society, Vol. LXXV., made by Dr. Fordyce, upon the weight said to be acquired by water upon being frozen; and being possessed of an excellent balance, belonging to his Most Serene Highness the Elector Palatine Duke of Bavaria; early in the beginning of the winter of the year 1787,—as soon as the cold was sufficiently intense for my purpose,—I set about to repeat those experiments, in order to convince myself whether the very extraordinary fact related might be depended on; and with a view to removing, as far as was in my power, every source of error and deception, I proceeded in the following manner.

Having provided a number of glass bottles, of the form and size of what in England is called a Florence flask,—blown as thin as possible,—and of the same shape and dimensions, I chose out from amongst them two, which, after using every method I could imagine of comparing them together, appeared to be so much alike as hardly to be distinguished from each other.

Into one of these bottles, which I shall call A, I put 4107.86 grains Troy of pure distilled water, which filled it about half full; and into the other, B, I put an equal weight of weak spirit of wine; and, sealing both the bottles hermetically, and washing them, and wiping them perfectly clean and dry on the outside, I suspended them to the arms of the balance, and placed the balance in a large room, which for some weeks had been regularly heated every day by a German stove, and in which the air was kept up to the temperature of 61° of Fahrenheit's thermometer, with very little variation. Having suffered the bottles, with their contents, to remain in this situation till I conceived they must have acquired the temperature of the circumambient air, I wiped them afresh, with a very clean, dry cambric handkerchief, and brought them into the most exact equilibrium possible, by attaching a small piece of very fine silver wire to the arm of the balance to which the bottle which was the lightest was suspended.

Having suffered the apparatus to remain in this situation about twelve hours longer, and finding no alteration in the relative weights of the bottles,—they continuing all this time to be in the most perfect equilibrium,—I now removed them into a large uninhabited room, fronting the north, in which the air, which was very quiet, was at the temperature of 29°F.; the air without doors being at the same time at 27°; and going out of the room, and locking the door after me, I suffered the bottles to remain forty-eight hours, undisturbed, in this cold situation, attached to the arms of the balance as before.

At the expiration of that time, I entered the room,—using the utmost caution not to disturb the balance,—when, to my great surprise, I found that the bottle A very sensibly preponderated.

The water which this bottle contained was completely frozen into one solid body of ice; but the spirit of wine, in the bottle B, showed no signs of freezing.

I now very cautiously restored the equilibrium by adding small pieces of the very fine wire of which gold lace is made, to the arm of the balance to which the bottle B was suspended, when I found that the bottle A had augmented its weight by $\frac{1}{35904}$ part of its whole weight at the beginning of the experiment; the weight of the bottle with its contents having been 4811.23 grains Troy (the bottle weighing 703.37 grains, and the water 4107.86 grains), and it requiring now $\frac{134}{1000}$ parts of a grain, added to the opposite arm of the balance, to counterbalance it.

Having had occasion, just at this time, to write to my friend, Sir Charles Blagden, upon another subject, I added a postscript to my letter, giving him a short account of this experiment, and telling him how "*very contrary to my expectation*" the result of it had turned out; but I soon after found that I had been too hasty in my communication. Sir Charles, in his answer to my letter, expressed doubts respecting the fact; but, before his letter had reached me, I had learned from my own experience how very dangerous it is in philosophical investigations to draw conclusions from single experiments.

Having removed the balance, with the two bottles attached to it, from the cold into the warm room (which still remained at the temperature of 61°), the ice in the bottle A gradually thawed; and, being at length totally reduced to water, and this water having acquired the temperature of the surrounding air, the two bottles, after being wiped perfectly clean and dry, were found to weigh as at the beginning of the experiment, before the water was frozen.

This experiment, being repeated, gave nearly the same result,—the water appearing when frozen to be heavier than in its fluid state; but some irregularity in the manner in which the water lost the additional weight which it had appeared to acquire upon being frozen when it was afterwards thawed, as also a sensible difference in the quantities of weight apparently acquired in the different experiments, led me to suspect that the experiment could not be depended on for deciding the fact in question. I therefore set about to repeat it, with some variations and improvements; but before I give an account of my further investigations relative to this subject, it may not be amiss to mention the method I pursued for discovering whether the appearances mentioned in the foregoing experiments might not arise from the imperfections of my balance; and it may likewise be proper to give an account, in this place, of an intermediate experiment which I made, with a view to discover, by a shorter route, and in a manner less exceptionable than that above mentioned, whether bodies actually lose or acquire any weight upon acquiring an additional quantity of latent heat.

My suspicions respecting the accuracy of the balance arose from a knowledge—which I acquired from the maker of it—of the manner in which it was constructed.

The three principal points of the balance having been determined, as nearly as possible, by measurement, the axes of motion were firmly fixed in their places, in a right line, and, the beam being afterwards finished, and its two arms brought to be in equilibrio, the balance was proved, by suspending weights, which before were known to be exactly equal, to the ends of its arms.

If with these weights the balance remained in equilibrio, it was considered as a proof that the beam was just; but if one arm was found to preponderate, the other was gradually lengthened, by beating it upon an anvil, until the difference of the lengths of the arms was reduced to nothing, or until equal weights, suspended to the two arms, remained in equilibrio; care being taken before each trial to bring the two ends of the beam to be in equilibrio, by reducing with a file the thickness of the arm which had been lengthened.

Though in this method of constructing balances the most perfect equality in the lengths of the arms may be obtained and consequently the greatest possible accuracy, when used at a time when the temperature of the air is the same as when the balance was made, yet, as it may happen that, in order to bring the arms of the balance to be of the same length, one of them may be much more hammered than the other, I suspected it might be possible that the texture of the metal forming the two arms might be rendered so far different by this operation as to occasion a difference in their expansions with heat; and that this difference might occasion a sensible error in the balance, when, being charged with a great weight, it should be exposed to a considerable change of temperature.

To determine whether the apparent augmentation of weight, in the experiments above related, arose in any degree from this cause, I had only to repeat the experiment, causing the two bottles A and B to change places upon the arms of the balance; but, as I had already found a sensible difference in the results of different repetitions of the same experiment, made as nearly as possible under the same circumstances, and as it was above all things of importance to ascertain the accuracy of my balance, I preferred making a particular experiment for that purpose.

My first idea was, to suspend to the arms of the balance, by very fine wires, two equal globes of glass, filled with mercury, and, suffering them to remain in my room till they should have acquired the known temperature of the air in it, to have removed them afterward into the cold, and to have seen if they still remained in equilibrio under such difference of temperature; but, considering the obstinacy with which moisture adheres to the surface of glass, and being afraid that somehow or other, notwithstanding all my precautions, one of the globes might acquire or retain more of it than the other, and that by that means its apparent weight might be increased; and having found by a former experiment, of which an account is given in one of the preceding papers (that on the Moisture absorbed from the Atmosphere by various Substances), that the gilt surfaces of metals do not attract moisture instead of the glass globes filled with mercury, I made use of two equal solid globes of brass, well gilt and burnished, which I suspended to the arms of the balance by fine gold wires.

These globes, which weighed 4975 grains each, being wiped perfectly clean, and having acquired the temperature (61°) of my room, in which they were exposed more than twenty-four hours, were brought into the most scrupulous equilibrium, and were then removed, attached to the arms of the balance, into a room in which the air was at the temperature of 26°, where they were left all night.

The result of this trial furnished the most satisfactory proof of the accuracy of the balance; for, upon entering the room, I found the equilibrium as perfect as at the beginning of the experiment.

Having thus removed my doubts respecting the accuracy of my balance, I now resumed my investigations relative to the augmentation of weight which fluids have been said to acquire upon being congealed.

In the experiments which I had made, I had, as I then imagined, guarded as much as possible against every source of error and deception. The bottles being of the same size, neither any occasional alteration in the pressure of the atmosphere during the experiment, nor the necessary and unavoidable difference in the densities of the air in the hot and in the cold rooms in which they were weighed, could affect their apparent weights; and their shapes and their quantities of surface being the same, and as they remained for such a considerable length of time in the heat and cold to which they were exposed, I flattered myself that the quantities of moisture remaining attached to their surfaces could not be so different as sensibly to affect the results of the experiments. But, in regard to this last circumstance, I afterwards found reason to conclude that my opinion was erroneous.

Admitting the fact stated by Dr. Fordyce,—and which my experiments had hitherto rather tended to corroborate than to contradict,—I could not conceive any other cause for the augmentation of the apparent weight of water upon its being frozen than the loss of so great a proportion of its latent heat as that fluid is known to evolve when it congeals; and I concluded that, if the loss of latent heat added to the weight of one body, it must of necessity produce the same effect on another, and consequently, that the augmentation of the quantity of latent heat must in all bodies and in all cases diminish their apparent weights.

To determine whether this is actually the case or not, I made the following experiment.

Having provided two bottles, as nearly alike as possible, and in all respects similar to those made use of in the experiments above mentioned, into one of them I put 4012.46 grains of water, and into the other an equal weight of mercury; and, sealing them hermetically, and suspending them to the arms of the balance, I suffered them to acquire the temperature of my room, 61°; then, bringing them into a perfect equilibrium with each other, I removed them into a room in which the air was at the temperature of 34°, where they remained twenty-four hours. But there was not the least appearance of either of them acquiring or losing any weight.

Here it is very certain that the quantity of heat lost by the water must have been very considerably greater than that lost by the mercury, the specific quantities of latent heat in water and in mercury having been determined to be to each other as 1000 to 33; but this difference in the quantities of heat lost produced no sensible difference on the weights of the fluids in question.

Had any difference of weight really existed, had it been no more than *one millionth* part of the weight of either of the fluids, I should certainly have discovered it; and had it amounted to so much as $\frac{1}{700000}$ part of that weight, I should have been able to have measured it, so sensible and so very accurate is the balance which I used in these experiments.

I was now much confirmed in my suspicions that the apparent augmentation of the weight of the water upon its being frozen, in the experiments before related, arose from some accidental cause; but I was not able to conceive what that cause could possibly be, unless it were either a greater quantity of moisture attached to the external surface of the bottle which contained the water than to the surface of that containing the spirits of wine, or some vertical current or currents of air caused by the bottles, or one of them not being exactly of the temperature of the surrounding atmosphere.

Though I had foreseen, and, as I thought, guarded sufficiently against, these accidents, by making use of bottles of the same size and form, and which were blown of the same kind of glass and at the same time, and by suffering the bottles in the experiments to remain for so considerable a length of time exposed to the different degrees of heat and of cold which alternately they were made to acquire; yet, as I did not know the relative conducting powers of ice and of spirit of wine with respect to heat, or, in other words, the degrees of facility or difficulty with which they acquire the temperature of the medium in which they are exposed, or the time taken up in that operation, and, consequently, was not *absolutely* certain as to the equality of the temperatures of the contents of the bottles at the time when their weights were compared, I determined now to repeat the experiments, with such variations as should put the matter in question out of all doubt.

I was the more anxious to assure myself of the real temperatures of the bottles and their contents, as any difference in their temperatures might vitiate the experiment, not only by causing unequal currents in the air, but also by causing, at the same time, a greater or less quantity of moisture to remain attached to the glass.

To remedy these evils, and also to render the experiment more striking and satisfactory in other respects, I proceeded in the following manner:—

Having provided three bottles, A, B, and C, as nearly alike as possible, and resembling in all respects those already described, into the first, A, I put 4214.28 grains of water, and a small thermometer, made on purpose for the experiment, and suspended in the bottle in such a manner that its bulb remained in the middle of the mass of water; into the second bottle, B, I put a like weight of spirit of wine, with a like thermometer; and, into the bottle C, I put an equal weight of mercury.

These bottles, being all hermetically sealed, were placed in a large room, in a corner far removed from the doors and windows, and where the air appeared to be perfectly quiet; and, being suffered to remain in this situation more than twenty-four hours, the heat of the room (61°) being kept up all that time with as little variation as possible, and the contents of the bottles A and B appearing, by their inclosed thermometers, to be exactly at the same temperature, the bottles were all wiped with a very clean, dry, cambric handkerchief; and, being afterwards suffered to remain exposed to the free air of the room a couple of hours longer, in order that any inequalities in the quantities of heat, or of the moisture attached to their surfaces, which might have been occasioned by the wiping, might be corrected by the operation of the atmosphere by which they were surrounded, they were all weighed, and were brought into the most exact equilibrium with each other, by means of small pieces of very fine silver wire, attached to the necks of those of the bottles which were the lightest.

This being done, the bottles were all removed into a room in which the air was at 30°, where they were suffered to remain, perfectly at rest and undisturbed, forty-eight hours; the bottles A and B being suspended to the arms of the balance, and the bottle C suspended, at an equal height, to the arm of a stand constructed for that purpose, and placed as near the balance as possible, and a very sensible thermometer suspended by the side of it.

At the end of forty-eight hours, during which time the apparatus was left in this situation, I entered the room, opening the door very gently for fear of disturbing the balance; when I had the pleasure to find the three thermometers, *viz.* that in the bottle A,—which was now inclosed in a solid cake of ice,—that in the bottle B, and that suspended in the open air of the room, all standing at the same point, 29°F., and the bottles A and B *remaining in the most perfect equilibrium.*

To assure myself that the play of the balance was free, I now approached it very gently, and caused it to vibrate; and I had the satisfaction to find, not only that it moved with the utmost freedom, but also, when its vibration ceased, that it rested precisely at the point from which it had set out.

I now removed the bottle B from the balance, and put the bottle C in its place; and I found that *that* likewise remained of the same apparent weight as at the beginning of the experiment, being in the same perfect equilibrium with the bottle A as at first.

I afterwards removed the whole apparatus into a warm room, and causing the ice in the bottle A to thaw, and suffering the three bottles to remain till they and their contents had acquired the exact temperature of the surrounding air, I wiped them very clean, and, comparing them together, I found their weights remained unaltered.

This experiment I afterwards repeated several times, and always with precisely the same result,—the water *in no instance* appearing to gain, or to lose, the least weight upon being frozen or upon being thawed; neither were the relative weights of the fluids in either of the other bottles in the least changed by the various degrees of heat and of cold to which they were exposed.

If the bottles were weighed at a time when their contents were not *precisely of the same temperature*, they would frequently appear to have gained, or to have lost, something of their weights; but this doubtless arose from the vertical currents which they caused in the atmosphere, upon being heated or cooled in it, or to unequal quantities of moisture attached to the surfaces of the bottles, or to both these causes operating together.

As I knew that the conducting power of mercury, with respect to heat, was considerably greater than either that of water or that of spirit of wine, while its capacity for receiving heat is much less than that of either of them, I did not think it necessary to inclose a thermometer in the bottle C, which contained the mercury; for it was evident that, when the contents of the other two bottles should appear, by their thermometers, to have arrived at the temperature of the medium in which they were exposed, the contents of the bottle C could not fail to have acquired it also, and even to have arrived at it before them; for the time taken up in the heating or in the cooling of any body, is, *caeteris paribus*, as the capacity of the body to receive and retain heat, *directly*, and its conducting power, *inversely*.

The bottles were suspended to the balance by silver wires about two inches long, with hooks at the ends of them; and, in removing and changing the bottles, I took care not to touch the glass. I likewise avoided upon all occasions, and particularly in the cold room, coming near the balance with my breath, or touching it, or any part of the apparatus, with my naked hands.

Having determined that water does not acquire or lose any weight upon being changed from a state of *fluidity* to that of *ice*, and *vice versa*, I shall now take my final leave of a subject which has long occupied me, and which has cost me much pains and trouble; being fully convinced, from the results of the above-mentioned experiments, that if heat be in fact a *substance*, or matter,—a fluid *sui generis*, as has been supposed,—which, passing from one body to another, and being accumulated, is the immediate cause of the phenomena we observe in heated bodies,—of which, however, I cannot help entertaining doubts,—it must be something so infinitely rare, even in its most condensed state, as to baffle all our attempts to discover its gravity. And if the opinion which has been adopted by many of our ablest philosophers, that heat is nothing more than an intestine vibratory motion of the constituent parts of heated bodies, should be well founded, it is clear that the weights of bodies can in no wise be affected by such motion.

It is, no doubt, upon the supposition that heat is a substance distinct from the heated body, and which is accumulated in it, that all the experiments which have been undertaken with a view to determine the weight which bodies have been supposed to gain or to lose upon being heated or cooled, have been made; and upon this supposition,—but without, however, adopting it entirely, as I do not conceive it to be sufficiently proved,—all my researches have been directed.

The experiments with *water* and with *ice* were made in a manner which I take to be perfectly unexceptionable, in which no foreign cause whatever could affect the results of them; and the quantity of heat which water is known to part with, upon being frozen, is so considerable, that if this loss has no effect upon its apparent weight, it may be presumed that we shall never be able to contrive an experiment by which we can render the weight of heat sensible.

Water, upon being frozen, has been found to lose a quantity of heat amounting to 140 degrees of Fahrenheit's thermometer; or—which is the same thing—the heat which a given quantity of water, previously cooled to the temperature of freezing, actually loses upon being changed to ice, if it were to be imbibed and retained by an equal quantity of water, at the given temperature (that of freezing), would heat it 140 degrees, or would raise it to the temperature of ($32° + 140$) 172° of Fahrenheit's thermometer, which is only 40° short of that of boiling water; consequently, any given quantity of water, at the temperature of freezing, upon being actually frozen, loses almost as much heat as, added to it, would be sufficient to make it boil.

It is clear, therefore, that the difference in the quantities of heat contained by the water in its fluid state and heated to the temperature of 61°F., and by the ice, in the experiments before mentioned, was very nearly equal to that between water in a state of boiling, and the same at the temperature of freezing.

But this quantity of heat will appear much more considerable when we consider the great capacity of water to contain heat, and the great apparent effect which the heat that water loses upon being frozen would produce were it to be imbibed by, or communicated to, any body whose power of receiving and retaining heat is much less.

The capacity of water to receive and retain heat—or what has been called its specific quantity of latent heat—has been found to be to that of gold as 1000 to 50, or as 20 to 1; consequently, the heat which any given quantity of water loses upon being frozen, were it to be communicated to an equal weight of gold at the temperature of freezing, the gold, instead of being heated 162 degrees, would be heated $140 \times 20 = 2800$ degrees, or, would be raised to a *bright red heat*.

It appears, therefore, to be clearly proved by my experiments, that a quantity of heat equal to that which 4214 grains (or about $9\frac{3}{4}$ oz.) of gold would require to heat it from the temperature of freezing water to be *red hot*, has no sensible effect upon a balance capable of indicating so small a variation of weight as that of $\frac{1}{1000000}$ part of the body in question; and, if the weight of gold is neither augmented nor lessened by *one millionth part*, upon being heated from the point of *freezing water* to that of a *bright red heat*, I think we may very safely conclude, that ALL ATTEMPTS TO DISCOVER ANY EFFECT OF HEAT UPON THE APPARENT WEIGHTS OF BODIES WILL BE FRUITLESS.

CHAPTER 7

WATER AS A NONCONDUCTOR OF HEAT

IN DEMONSTRATING heat conduction by convection currents, Rumford proved that there was a mechanism in a fluid which provided a mechanical means for transmitting heat. In the following paper he attempted to demonstrate that except for the mass motion of particles, water was an insulator. Although we know today that this is not strictly true, within the accuracy of his measurements he showed that water did not conduct heat as long as he arranged his experimental conditions to eliminate convection flow.

These series of experiments convinced Rumford that conduction of heat through a fluid was entirely the result of mass motion of the fluid itself, and, therefore, in demonstrating the requirement for a mechanical motion for the transmission of heat, he eliminated the caloricists' explanation that heat was carried from one place to another by the motion of the subtle fluid called caloric.

In these long studies of the mechanism of heat flow through and by means of liquids, the anticaloricists were demonstrating that those fluids which were available to them for experimentation were not themselves conductors of heat, and by implication the fluid caloric might itself have such properties as to weaken the strength of argument on the side of the caloric theory of heat.

Inquiries concerning the Mode of the Propagation of Heat in Liquids

Read before the National Institute of France
June 9, 1806

Bibliotheque Britannique (Science et Arts) XXXII, 123–141 (1806)
Nicholson's Journal XIV, 353–363 (1806)

The motions in fluids which result from a change in their temperature give rise to so great a number of phenomena, that philosophers cannot bestow too much pains in investigating that interesting branch of knowledge.

When heat is propagated in solid bodies, it passes from particle to particle, *de proche en proche*, and apparently with the same celerity in every direction; but it is certain that heat is not transmitted in the same manner in fluids.

When a solid body is heated and plunged in a cold liquid, the particles of the liquid in contact with the body, being rarefied by the heat that they receive from it, and being rendered specifically lighter than the surrounding particles, are forced to give place to these last and to rise to the surface of the liquid; and the cold particles that replace them at the surface of the hot body, being in their turn heated, rarefied, and forced up,—all the particles thus heated by a successive contact with the hot body form a continued ascending current, which carries the whole of the heat immediately towards the surface of the liquid, so that the strata of the liquid situated at a small distance under the hot body are not sensibly heated by it.

When a solid body is plunged in a liquid which is hotter than the body, the particles of the liquid in contact with the body, being condensed by the cooling they undergo, descend, in consequence of the increase of their specific gravity, and fall to the bottom of the liquid; and the strata situated above the level of the cold body are not cooled by it immediately.

It is true that the viscosity of liquids, even of those which possess the highest known degree of fluidity, is still much too great to allow one of their particles individually being moved out of its place by any change of specific gravity occasioned by heat or cold; yet this does not prevent currents from being formed, in the manner above described, by small masses of the liquid composed of a great number of such particles.

The existence of currents in the ordinary cases of the heating and cooling of liquids cannot any longer be called in question; but philosophers are not yet agreed with respect to the extent of the effects produced by those currents.

In treating of abstruse subjects, it is indispensably necessary to fix with precision the exact meaning of the words we employ. The distinction established between *conductors* and *non-conductors* of heat is too vague not to stand in need of explanation. An example will show the ambiguity of these expressions.

If two equal cubes of any solid matter,—copper, for instance,—of two inches in diameter, the one at the temperature of 60°, the other at 100°, be placed one above the other, the cold cube will be heated by the hot one, and this last will be cooled.

If the cold cube be placed upon a table and its upper surface covered by a large plate of metal,—of silver, for instance,—a quarter of an inch thick, and if the hot cube be placed upon this plate immediately above the cold cube, the heat will descend through the metallic place with a certain degree of facility, and will heat the cold cube.

If a dry board of the same thickness with the metallic plate be substituted in its place, the heat will descend through the wood, but with much less celerity than through the plate of silver.

But if a stratum of water or of any other liquid be substituted in place of the metallic plate or of the board, the result will be very different. If, for instance, the cold cube being placed in a large tub resting on the middle of its bottom, the hot cube be suspended over it by cords, or in any other manner so that the lower surface of the hot cube be immediately above the upper surface of the cold cube, at the distance of a quarter of an inch, and the tub be then filled with water at the same temperature as that of the cold cube, the heat will not descend from the hot cube to the cold one through the stratum of water of a quarter of an inch in thickness that separates them.

We may with propriety call silver a *good conductor* of heat, and dry wood a *bad conductor*; but what shall we say of water? I have called it a *non-conductor* for want of a more suitable term, but I always felt that that word expresses but imperfectly the quality that was meant to be designated.

In the experiment of the two cubes plunged in water, if the hot cube be placed below and the cold cube above it, the heat will not only be communicated from the hot to the cold cube, but it will pass even more rapidly than when the two cubes are separated by a plate of silver. But in this case it is evident that the heat is *transported* by the ascending currents which are formed in the liquid in consequence of the heat which it receives from the hot body.

The existence of these currents in certain cases has been known a long time, but philosophers have not been sufficiently attentive to the many curious phenomena that depend upon them. It has not even been suspected with what extreme slowness heat passes in fluids, from particle to particle, *de proche en proche*, in cases where the effects of such communication becomes sensible.

For some time after I had engaged in this interesting inquiry, I conceived that this kind of communication was absolutely impossible in all cases; but a more attentive examination of the phenomena has convinced me that this conclusion was too hasty. As early as the beginning of 1800, in a note published in the third edition of my Seventh Essay, I announced a conjecture that the non-conducting power of fluids might perhaps depend solely on the extreme mobility of their particles; and it is certain, if this conjecture is well founded, liquids must necessarily become conductors of heat (though very imperfect ones) in all cases where this mobility of their particles is destroyed, as well as in these rare but yet possible cases, where a change of temperature can take place in a liquid without giving its particles any tendency to move, or to be moved out of their places.

The unequivocal results of a great many experiments have shown, that in ordinary cases, and perhaps in all cases where heat is propagated in considerable masses of fluids, its distribution is accomplished precisely in the manner that the new theory supposes, that is to say, by currents. And it is certain that the knowledge of that fact has enabled us to explain in a satisfactory manner several interesting phenomena of nature, which before were enveloped in much obscurity.

When a hot solid body is plunged in a cold liquid, there can be no doubt concerning the existence of the vertical ascending currents which are formed in the liquid, and which convey to the surface the heat which its particles have received; but with respect to the strata of liquid situated under the hot body, *are they or are they not heated by this body by means of a direct communication of heat from above downwards, from particle to particle, these particles remaining in their places*? This is a question on which philosophers are not yet agreed. As it is a question of great importance, I have long meditated on the means of deciding it; and after several unsuccessful attempts, I have at last succeeded in making an experiment which I think is decisive.

As the apparatus which I used for this experiment, and which I

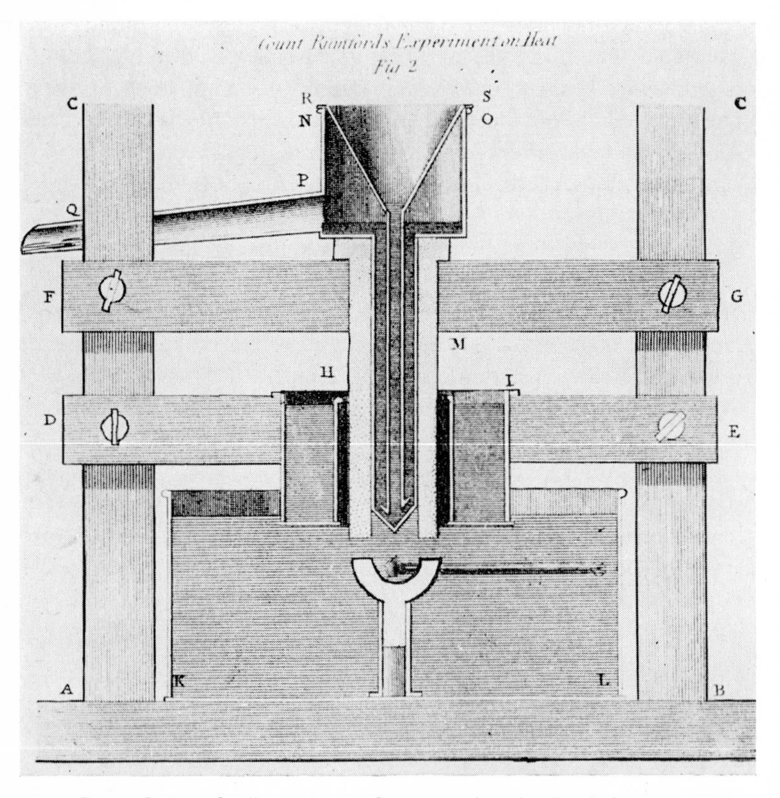

PLATE 8. Rumford's apparatus for measuring the thermal conductivity of water.

have the honour of laying before the assembly, is somewhat complicated; and as it is indispensably necessary to be intimately acquainted with it, in order to form a judgment concerning the degree of confidence which the results of the experiment may deserve,—it is necessary to give a detailed description of this machinery. The annexed figure gives a distinct representation of its principal parts. It is drawn on a scale of a quarter of an inch to the inch, English measure.

A B [Plate 8] is a board, of oak, seen in profile; it is 1½ inches thick, 18 inches long, and 11 inches in breadth. It serves to support two square upright pillars, C C, 18½ inches in height and 1½ inches square. They are firmly fixed in the board at the distance of 11 inches asunder, and serve to support the two cross-pieces, D E, F G, at different heights.

These cross-pieces are each pierced with two square holes, at the distance of 11 inches one from the other, into which the upright pillars C C enter, and the cross-pieces are supported at any height that is required, by means of a screw of compression. These screws are represented in the figure.

The cross-piece F G, which is represented in profile, is 17 inches in length, and 1½ inches thick, and 3 inches in breadth. It is pierced in the middle by a cylindrical hole of 2 inches in diameter.

The cross-piece D E is 17 inches in length by 1½ inches in thickness. It is 3 inches wide at each end and 6 inches in the middle, where it is pierced by a circular hole 5 inches in diameter.

The cross-piece D E serves to support the annular vessel H I, of which a vertical section passing through its axis is seen in the figure. This vessel, formed of thin brass plates, is 5 inches in diameter without, 3 inches in diameter within, and 27⅛ inches in depth. This vessel is filled with water during the experiments to the height of 2½ inches; and its form is such, that, if the water that it contains were frozen into a solid mass of ice, this piece of ice would have the form of a tube or perforated cylinder of 1 inch in thickness and 2½ inches high by 5 inches in diameter without. Its cylindrical cavity would be precisely 3 inches in diameter.

K L is a vertical and central section of a cylindrical vessel of tin of 10 inches in diameter by $4\frac{1}{2}$ inches in depth. It is filled with water to the height of 4 inches, as it is seen in the figure.

The cross-piece D E is placed at such a height that the bottom of the annular vessel H I is plunged a quarter of an inch under the surface of the water contained in the great cylindrical vessel K L.

In the axis of this last vessel is placed a small hemispherical cup of wood 2 inches in diameter without and $\frac{1}{2}$ of an inch thick. It is kept in its place by a short vertical tube of tin, soldered to the bottom of the cylindrical vessel K L, into which the stalk of the cup fits tightly.

The middle of the cavity of this cup is occupied by the bulb of a small mercurial thermometer of great sensibility. Its tube, which has an ivory scale, is laid down horizontally, and fixed in one side of the cup, through which the tube passes, in such a manner that the lowest part of the bulb is elevated $\frac{1}{10}$ of an inch above the bottom of the cup. The diameter of the bulb being $\frac{3}{10}$ of an inch, and the hemispherical cup having $\frac{1}{2}$ inch of radius within, it is evident that the upper part of the bulb is $\frac{1}{10}$ of an inch below the level of the brim of the cup that contains it. To avoid charging the figure with too many details, the scale of the thermometer is not drawn, but the tube is distinctly represented.

The horizontal cross-piece F G serves to support a very essential part of the apparatus, which remains to be described.

This cross-piece supports, in the first place, a vertical tube of wood, M, $6\frac{6}{10}$ inches in length and 2 inches in diameter without. Its interior diameter is $1\frac{1}{20}$ inch. This tube is supported by a projecting collar (represented in the figure), $2\frac{1}{2}$ inches in diameter, which rests on the cross-piece F G. It is a vertical and central section of this tube that is represented in the figure, and it is dotted in order to distinguish it from the surrounding parts of the apparatus.

The lower part of this tube is plunged $\frac{6}{10}$ of an inch under the surface of the water in the large cylindrical vessel K L; and it is placed precisely above the wooden cup in the prolongation of its axis, the lower extremity of the tube being at the distance of $\frac{3}{10}$ of an inch above the horizontal level of the brim of the cup.

On the top of the tube of wood is placed a cylindrical vessel

N O, of sheet brass, 3 inches in diameter, $2\frac{3}{4}$ inches high, which has a lateral spout, P Q, placed a little above the level of its bottom.

From the middle of the bottom of this vessel, there descends a cylindrical tube of brass, 6 inches in length and 1 inch in diameter, which ends below in a hollow conical point, as represented in the figure.

R S is a vertical and central section of a funnel of brass, which ends below in a cylindrical tube of $\frac{3}{10}$ of an inch in diameter and $6\frac{6}{10}$ inches long. This funnel is kept in its place in the axis of the cylindrical vessel N O by the exact fitting of its upper edge upon that of the vessel into which it is adjusted.

The lower end of the tube of this funnel is surrounded by a projecting edge or flange in the form of a hollow inverted cone. The diameter of this conical projecting brim above, at its base, is $\frac{7}{10}$ of an inch, and it is soldered below to the end of the tube.

When hot water is poured into the funnel, this liquid, descending by the tube of the funnel, strikes against the inner surface of the hollow inverted cone which terminates the vertical tube that belongs to the vessel N O, and then, rising up through this last tube into that vessel, it runs off by its spout. It was with a view to force this water to come into more intimate contact with the hollow cone that the projecting edge, in form of an inverted cone, was added to the lower end of the tube of the funnel.

The object chiefly in view in the arrangement of this apparatus was to give to the conical point which terminates the vertical tube of the vessel N O, an elevated temperature, which should remain constant during some time, for the purpose of observing if the heat, which must necessarily be communicated by this metallic point to the small quantity of water with which it is in contact, and which is confined in the lower part of the wooden tube M, would descend, or not, to the thermometer which was placed in the wooden cup.

There was still one source of error and uncertainty against which it was necessary to guard. The heat communicated through the sides of the wooden tube to the water contained in the great cylindrical vessel K L might be transported to the sides of that vessel, and, being then communicated from above downwards through these sides, might heat successively the lower strata of the liquid, and at last that stratum in which the thermometer was.

It was to prevent this that the annular vessel H I was used, and it performed its office in the following manner: The particles of water contained in the great vessel K L, which, being in contact with the exterior surface of the wooden tube, were heated by that tube, could not fail to rise to the surface, and there they necessarily came into contact with the interior sides of the annular vessel, to which they communicated the excess of heat they had received from the wooden tube.

This heat, passing readily through the thin metallic sides of that vessel, was given off as fast as it was received to the particles of cold water contained in the vessel which were in contact with its sides, and these particles, rising to the surface of the water contained in the annular vessel in consequence of their acquired heat and levity, the progress of the heat from the wooden tube to the sides of the large vessel K L, was interrupted, and all the heat that passed through the sides of the wooden tube was by these means turned aside in such a manner that it could no longer disturb the progress of the experiment, nor affect the certainty of its results.

Before I proceed to give an account of the result of this inquiry, I shall take the liberty to recall the attention of the Assembly to the most important circumstances of the experiment.

On pouring boiling water in a small uninterrupted stream into the funnel, the hollow conical point which terminates the vertical tube belonging to the vessel N O was heated, and kept at a constant temperature little under that of boiling water.

This point was surrounded by a small quantity of water contained in the cavity of the lower part of the wooden tube, and as this water could not change its place nor be displaced by the surrounding cold water, being enclosed and protected by the sides of the wooden tube, it would necessarily become very hot in a short time.

But this small quantity of hot water lay immediately upon a stratum of cold water, which separated it from the bulb of the thermometer, placed directly under it at the distance of only half an inch.

If heat could pass in the water from above downwards, it would no doubt pass from the lower stratum of hot water contained in the open end of the wooden tube to the bulb of the thermometer, which lay immediately below it and at so small a distance.

Three experiments were made with this apparatus, and always with exactly the same results. In the first, a stream of boiling water was poured into the funnel during 10 minutes; in the second, during 12 minutes; and in the third, during 15 minutes.

The thermometer, whose bulb was in the wooden cup, remained *at perfect rest* from the beginning of the experiment to the end of it without showing the slightest sign of being in any way affected by the hot water which was so near it.

These experiments were made at Munich in the month of July, 1805; the temperature of the air and of the water contained in the vessel K L being 70° Fahrenheit.

A small thermometer placed in the water contained in the annular vessel H I, in such a manner that its bulb was scarcely submerged, marked that this water had received a little heat in each of the three experiments.

Another similar thermometer placed in the water contained in the large vessel K L, immediately under its surface and near one side of the vessel, showed that this water had not acquired any sensible increase of temperature during the experiments.

From the results of these experiments we are authorized to conclude, that heat does not descend in water to a sensible distance, in cases where the particles of the liquid which receive heat are exposed to be displaced and forced upwards by the surrounding colder and denser particles, that is to say, in all the cases (and they are the most common) where heat is applied to the strata of the liquid situated under its surface.

But the results of the experiments in question do not prove that heat cannot in any case descend in water; and still less can it be inferred from them, that all direct communication of heat in this liquid, from particle to particle, *de proche en proche*, is impossible.

They do not even prove that heat did not descend, *to a small distance*, below the level of the end of the wooden tube in these experiments; for it is certain that that event could take place without the thermometer, which was situated a little lower, being in any way affected by that heat.

The particles of water situated at a very small distance below the level of the lower end of the wooden tube, being heated by the stratum of hot water which rested immediately on them, might have been displaced by the surrounding colder and denser particles, and forced to rise to the surface; and these last being in their turn heated, forced upwards and replaced by other cold particles, it is evident that the heat could not make its way downwards so far as to arrive at the thermometer through a stratum of liquid, which, though apparently at rest, was nevertheless in part composed of particles which were continually changing.

I have long suspected that the apparent impossibility of a direct communication of heat between neighbouring particles of fluids depends solely on the great mobility of those particles (see note, p. 202, Vol. II. of my Essays, 3d edition, London, 1800); and if this suspicion be well founded, it is certain that when this mobility ceases, the effect which depends on it must cease likewise.

* * *

The inquiry as to the non-conducting power of fluids—an inquiry to which my experiments and observations have given rise—is, no doubt, of great interest to science; and, whatever may be the final result of its investigation, I shall regard myself fortunate in having drawn the attention of a great number of enlightened philosophers towards an object which was long negected, and which was so worthy of being studied.

CHAPTER 8

PROPAGATION OF HEAT IN VARIOUS SUBSTANCES

It is difficult to get the real flavor of late eighteenth-century scientific writing without exposing oneself to a fairly lengthy sample of this kind of journalism. One must read the following paper with the idea that not only is it a scientific report in which a clearly defined set of relevant parameters were not yet established, but also it was written much more in the vein of a popular essay than would ever be tolerated in our strictly scientific journals of today including the *Philosophical Transactions* of the Royal Society in which it appeared in 1786 and 1787.

Thus besides being a report of considerable scientific merit, it includes anecdotes of exploding apparatus, the people who assisted and watched the demonstrations; facts which now appear to us irrelevant, such as the date, the weather, the barometric pressure, the relative humidity, the air temperature, and so on, which at that time were not so out of place because Rumford did not know whether these facts were relevant to his experiments or not, as well as some rather delightful editorial comments. It would be inconceivable today to have the reviewers of a scientific article allow such comments as "The snows which cover the surface of the earth in winter, in high latitudes, are doubtless designed by an all-provident Creator as a garment to defend it against the piercing winds . . ." or to assign the reason for a set of observations to the "infinite wisdom and goodness of Divine Providence guarding us against evil effects". One might question, however, whether something has not been lost in the drab efficiency of our modern scientific writing.

ESSAY VIII

Of the Propagation of Heat in various substances

Philosophical Transactions LXXVI, 273–304 (1786)
Philosophical Transactions LXXVII 48–80 (1787)
Published separately, London 1798

Examining the conducting power of air, and of various other fluid and solid bodies, with regard to Heat, I was led to examine the conducting power of the *Torricellian vacuum*. From the striking analogy between the electric fluid and Heat respecting their conductors and non-conductors (having found that bodies, in general, which are conductors of the electric fluid, are likewise good conductors of Heat, and, on the contrary, that electric bodies, or such as are bad conductors of the electric fluid, are likewise bad conductors of Heat), I was led to imagine that the Torricellian vacuum, which is known to afford so ready a passage to the electric fluid, would also have afforded a ready passage to Heat.

The common experiments of heating and cooling bodies under the receiver of an air-pump I conceive to be inadequate to determining this question; not only on account of the impossibility of making a perfect void of air by means of the pump, but also on account of the moist vapour, which, exhaling from the wet leather and the oil used in the machine, expands under the receiver, and fills it with a watery fluid, which, though extremely rare, is yet capable of conducting a great deal of Heat: I had recourse, therefore, to other contrivances.

I took a thermometer, unfilled, the diameter of whose bulb (which was globular) was just half an inch, Paris measure, and fixed it in the center of a hollow glass ball of the diameter of $1\frac{3}{4}$ Paris inch, in such a manner that, the short neck or opening of the ball being soldered fast to the tube of the thermometer $7\frac{1}{2}$ lines above its bulb, the bulb of the thermometer remained fixed in the center of the ball, and consequently was cut off from all communication with the external air. In the bottom of the glass ball was fixed a small hollow tube or point, which projecting outwards was soldered to the end of a common barometer tube about 32 inches in length, and by means of this opening the space between the internal surface of the glass ball and the bulb of the thermometer was filled with hot mercury, which had been previously freed of air and moisture by boiling. The ball, and also the barometrical tube attached to it, being filled with mercury, the tube was carefully inverted, and its open end placed in a bowl in which there was a quantity of mercury. The instrument now became a barometer, and the mercury descending from the ball (which was now uppermost) left the space surrounding the bulb of the thermometer free of air. The mercury having totally quitted the glass ball, and having sunk in the tube to the height of 28 inches (being the height of the mercury in the common barometer at that time), with a lamp and a blow-pipe I melted the tube together, or sealed it hermetically, about three quarters of an inch below the ball, and, cutting it at this place with a fine file, I separated the ball from the long barometrical tube. The thermometer being afterwards filled with mercury in the common way, I now possessed a thermometer whose bulb was confined in the center of a *Torricellian vacuum*, and which served at the same time as the body to be heated, and as the instrument for measuring the Heat communicated.

Experiment No. 1

With this instrument (see Fig. 1 [Plate 9]) I made the following experiment. Having plunged it into a vessel filled with water, warm to the 18th degree of Reaumur's scale, and suffered it to remain there till it had acquired the temperature of the water, that is to say, till the mercury in the inclosed thermometer stood at 18°, I took it out of this vessel and plunged it suddenly into a vessel of boiling water, and holding it in the water (which was kept constantly boiling) by the end of the tube, in such a manner that the glass ball, in the center of which was the bulb of the thermometer, was just submerged, I observed the number of degrees to which the mercury in the thermometer had arisen at different periods of time, counted from the moment of its immersion. Thus, after it had remained in the boiling water 1 min. 30 sec. I found the mercury had risen from 18° to 27°. After 4 minutes had elapsed, it had risen $44\frac{9}{10}$°; and at the end of 5 minutes it had risen to $48\frac{2}{10}$°.

Experiment No. 2

Taking it now out of the boiling water I suffered it to cool gradually in the air, and after it had acquired the temperature of the atmosphere, which was that of 15° R. (the weather being perfectly fine), I broke off a little piece from the point of the small tube which remained at the bottom of the glass ball, where it had been hermetically sealed, and of course the atmospheric air rushed immediately into the ball. The ball surrounding the bulb of the thermometer being now filled with air (instead of being emptied of air, as it was in the before-mentioned experiment), I resealed the end of the small tube at the bottom of the glass ball hermetically, and by that means cut off all communication between the air confined in the ball and the external air; and with the instrument so prepared I repeated the experiment before mentioned, that is to say, I put it into water warmed to 18°, and when it had acquired the temperature of the water, I plunged it into boiling water, and observed the times of the ascent of the mercury in the thermometer. They were as in Table [1]:—

TABLE 1

	Time elapsed	Heat acquired
Heat at the moment of being plunged into the boiling water		18° R.
	m. s.	0
After having remained in the boiling water	0 45	27
	1 0	$34\frac{4}{10}$
	2 10	$44\frac{9}{10}$
	2 40	$48\frac{2}{10}$
	4 0	$56\frac{2}{10}$
	5 0	$60\frac{9}{10}$

From the result of these experiments it appears, evidently, that the Torricellian vacuum, which affords so ready a passage to the electric fluid, so far from being a good conductor of Heat, is a much worse conductor of it than common air, which of itself is reckoned among the worst; for in the last experiment, when the bulb of the thermometer was surrounded with air, and the instrument was plunged into boiling water, the mercury rose from 18° to 27° in 45 seconds; but in the former experiment, when it was surrounded by a Torricellian vacuum, it required to remain in the boiling water 1 minute 30 seconds = 90 seconds, to acquire that degree of heat. In the vacuum it required 5 minutes to rise to $48\frac{2}{10}$°; but in air it rose to that height in 2 minutes 40 seconds; and the proportion of the times in the other observations is nearly the same, as will appear in Table [2].

TABLE 2

	The bulb of the thermometer placed in the center of the glass ball, and					
	surrounded by a Torricellian vacuum (Exp. No. 1)			surrounded by air (Exp. No. 2)		
	Time elapsed		Heat acquired	Time elapsed		Heat acquired
Upon being plunged into boiling water			18°			18°
	m.	s.	0	m.	s.	0
After remaining in it	1	30	27	0	45	27
	—		—	1	0	$30\frac{4}{10}$
	4	0	$44\frac{9}{10}$	2	10	$44\frac{9}{10}$
	5	0	$48\frac{2}{10}$	2	40	$48\frac{2}{10}$
	—		—	4	0	$56\frac{2}{10}$
	—		—	5	0	$60\frac{9}{10}$

These experiments were made at Manheim, upon the first day of July, 1785, in the presence of Professor Hemmer, of the Electoral Academy of Sciences of Manheim, and Charles Artaria meteorological instrument maker to the Academy, by whom I was assisted in making them.

Finding the construction of the instrument made use of in these experiments attended with much trouble and risk, on account of the difficulty of soldering the glass ball to the tube of the thermometer without at the same time either closing up, or otherwise injuring, the bore of the tube, I had recourse to another contrivance much more commodious, and much easier in the execution.

At the end of a glass tube or cylinder about eleven inches in length, and near three quarters of an inch in diameter internally, I caused a hollow globe to be blown $1\frac{1}{2}$ inch in diameter, with an opening in the bottom of it corresponding with the bore of the tube, and equal to it in diameter, leaving to the opening a neck or short tube, about an inch in length. Having a thermometer prepared, whose bulb was just half an inch in diameter, and whose freezing point fell at about $2\frac{3}{4}$ inches above its bulb, I graduated its tube according to Reaumur's scale, beginning at 0°, and marking that point, and also every tenth degree above it to 80°, with threads of fine silk bound round it, which being moistened with lac varnish adhered firmly to the tube. This thermometer I introduced into the glass cylinder and globe just described, by the opening in the bottom of the globe, having first choaked the cylinder at about 2 inches from its junction with the globe by heating it, and crowding its sides inwards towards its axis, leaving only an opening sufficient to admit the tube of the thermometer. The thermometer being introduced into the cylinder in such a manner that the center of its bulb coincided with the center of the globe, I marked a place in the cylinder, about three quarters of an inch above the 80th degree or boiling point upon the tube of the inclosed thermometer, and taking out the thermometer, I choaked the cylinder again in this place. Introducing now the thermometer for the last time, I closed the opening at the bottom of the globe at the lamp, taking care before I brought it to the fire, to turn the cylinder upside down, and to let the bulb of the thermometer fall into the cylinder till it rested upon the lower choak in the cylinder. By this means the bulb of the thermometer was removed more than 3 inches from the flame of the lamp. The opening at the bottom of the globe being now closed, and the bulb of the thermometer being suffered to return into the globe, the end of the cylinder was cut off to within about half an inch of the upper choak. This being done, it is plain that the tube of the thermometer projected beyond the end of the cylinder. Taking hold of the end of the tube, I placed the bulb of the thermometer as nearly as possible in the center of the globe, and observing and marking a point in the tube immediately above the upper choak of the cylinder, I turned the cylinder upside down, and suffering the bulb of the thermometer to enter the cylinder, and rest upon the first or lower choak, (by which means the end of the tube of the thermometer came further out of the cylinder,) the end of the tube was cut off at the mark just mentioned, (care having first been taken to melt the internal cavity or bore of the tube together at that place,) and a small solid ball of glass, a little larger than the internal diameter or opening of the choak, was soldered to the end of the tube, forming a little button or knob, which resting upon the upper choak of the cylinder served to suspend the thermometer in such a manner that the center of its bulb coincided with the center of the globe in which it was shut up. The end of the cylinder above the upper choak being now heated and drawn out to a point, or rather being formed into the figure of the frustum of a hollow cone, the end of it was soldered to the end of a barometrical tube, by the help of which the cavity of the cylinder and globe containing the thermometer was completely voided of air with mercury; when, the end of the cylinder being hermetically sealed, the barometrical tube was detached from it with a file, and the thermometer was left completely shut up in a Torricellian vacuum, the centre of the bulb of the thermometer being confined in the centre of the glass globe, without touching it in any part, by means of the two choaks in the cylinder, and the button upon the end of the tube. (See Fig. 2 [Plate 9].)

Of these instruments I provided myself with two, as nearly as possible of the same dimensions; the one, which I shall call No. 1, being voided of air, in the manner above described; the other, No. 2, being filled with air, and hermetically sealed.

With these two instruments (see Fig. 2 [Plate 9]) I made the following experiments upon the 11th of July last at Manheim, between the hours of ten and twelve, the weather being very fine and clear, the mercury in the barometer standing at 27 inches 11 lines, Reaumur's thermometer at 15°, and the quill hygrometer of the Academy of Manheim at 47°.

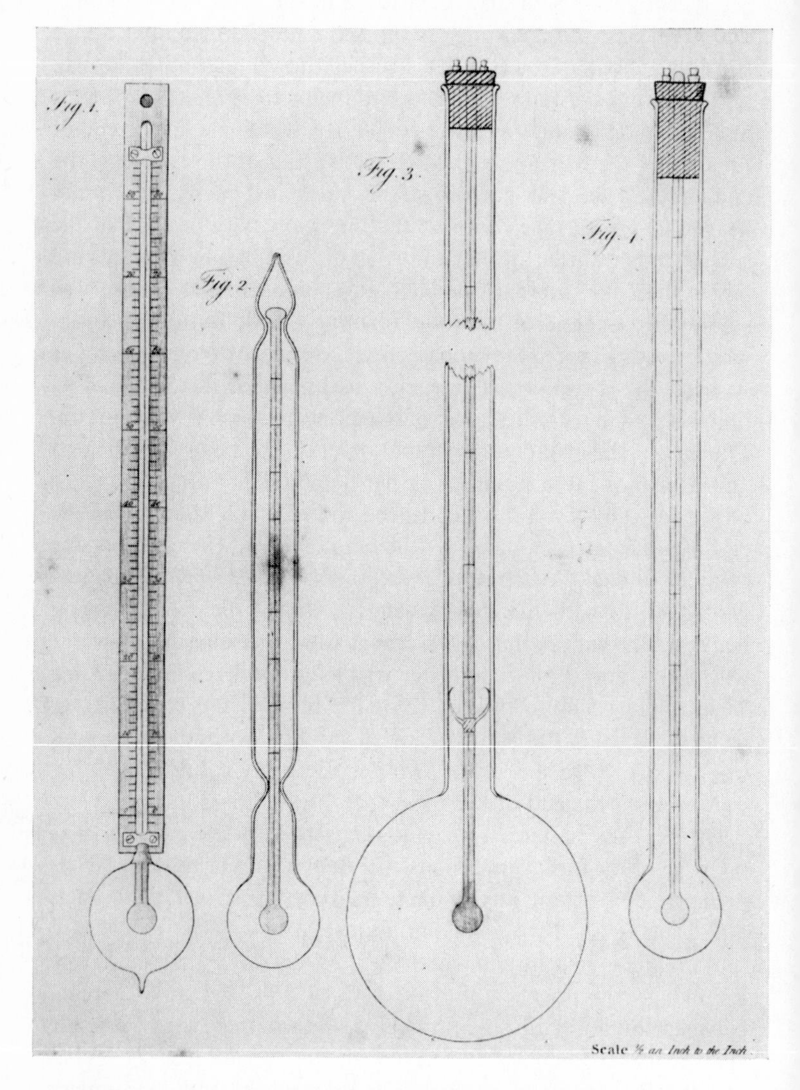

PLATE 9. Thermometers used in the studies of the propagation of heat in various substances.

Experiments No. 3, 4, 5, and 6

Putting both the instruments into a mixture of pounded ice and water, I let them remain there till the mercury in the inclosed thermometers rested at the point 0°, that is to say, till they had acquired exactly the temperature of the cold mixture; and then taking them out of it I plunged them suddenly into a large vessel of boiling water, and observed the time required for the mercury to rise in the thermometers from ten degrees to ten degrees, from

TABLE 3

Thermometer No. 1	*Thermometer No.* 2
Its bulb half an inch in diameter, shut up in the center of a hollow glass globe, $1\frac{1}{2}$ inch in diameter, *void of air*, and hermetically sealed.	Its bulb half an inch in diameter, shut up in the center of a hollow glass globe, $1\frac{1}{2}$ inch in diameter, *filled with air*, and hermetically sealed.
Taken out of freezing water, and plunged into boiling water.	*Taken out of freezing water, and plunged into boiling water.*

Time elapsed		Heat acquired	Time elapsed		Heat acquired
Exp. No. 3	Exp. No. 4		Exp. No. 5	Exp. No. 6	
		0°			0°
M. S.	M. S.	0	M. S.	M. S.	0
0 51	0 51	10	0 30	0 30	10
0 59	0 59	20	0 35	0 37	20
1 1	1 2	30	0 41	0 41	30
1 18	1 22	40	0 49	0 53	40
1 24	1 23	50	1 1	0 59	50
2 0	1 51	60	1 24	1 20	60
3 30	3 6	70	2 45	2 25	70
11 41	10 27	80	9 10	9 38	80
22 44	21 1	= total time of heating from 0° to 80°	16 55	17 3	= total time of heating from 0° to 80°

Thermometer No. 1		Thermometer No. 2	
Total time from 0° to 70°:	M. S.	Total time from 0° to 70°:	M. S.
In Exp. No. 3 =	11 3	In Exp. No. 5 =	7 45
In Exp. No. 4 =	10 34	In Exp. No. 6 =	7 25
Medium =	10 $48\frac{1}{2}$	Medium =	7 35

0° to 80°, taking care to keep the water constantly boiling during the whole of this time, and taking care also to keep the instruments immersed to the same depth, that is to say, just so deep that the point 0° of the inclosed thermometer was even with the surface of the water.

These experiments I repeated twice with the utmost care; and Table [3] gives the result of them.

It appears from these experiments that the conducting power of air to that of the Torricellian vacuum, under the circumstances described, is as $7\frac{35}{60}$ to $10\frac{48\frac{1}{2}}{60}$ inversely, or as 1000 to 702 nearly; for, the quantities of Heat communicated being equal, the intensity of the communication is as the times inversely.

TABLE 4

Thermometer No. 1 Surrounded by a Torricellian vacuum *Taken out of boiling water, and plunged into freezing water.*			*Thermometer No. 2* Surrounded by air. *Taken out of boiling water, and plunged into freezing water.*		
Time elapsed		Heat lost	Time elapsed		Heat lost
Exp. No. 7	Exp. No. 8		Exp. No. 9	Exp. No. 10	
		80°			80°
M. S.	M. S.	0	M. S.	M. S.	0
1 2	0 54	70	0 33	0 33	70
0 58	1 2	60	0 39	0 34	60
1 17	1 18	50	0 44	0 44	50
1 46	1 37	40	0 55	0 55	40
2 5	2 16	30	1 17	1 18	30
3 14	3 10	20	1 57	1 57	20
5 42	5 59	10	3 44	3 40	10
Not observed.	Not observed.	0	40 10	Not observed.	0

Total time of cooling from 80° to 10°	M. S.	Total time of cooling from 80° to 10°	M. S.
In Exp. No. 7=	16 4	In Exp. No. 9=	9 49
In Exp. No. 8=	16 16	In Exp. No. 10=	9 41
Medium=	16 10	Medium=	9 45

In these experiments the Heat passed through the surrounding medium *into* the bulb of the thermometer: in order to reverse the experiment, and make the Heat pass *out of* the thermometer, I put the instruments into boiling water, and let them remain therein till they had acquired the temperature of the water, that is to say, till the mercury in the inclosed thermometers stood at 80°; and then, taking them out of the boiling water, I plunged them suddenly into a mixture of water and pounded ice, and moving them about continually in this mixture, I observed the times employed in cooling as in Table [4].

By these experiments it appears that the conducting power of air is to that of the Torricellian vacuum as $9\frac{45}{60}$ to $16\frac{10}{60}$ inversely, or as 1000 to 603.

To determine whether the same law would hold good when the heated thermometers, instead of being plunged into freezing water, were suffered to cool in the open air, I made the following experiments. The thermometers No. 1 and No. 2 being again heated in boiling water, as in the last experiments, I took them out of the water, and suspended them in the middle of a large room, where

TABLE 5

(Exp. No. 11) *Thermometer No.* 1 Surrounded by a Torricellian vacuum. *Heated to* 80°, *and suspended in the open air warm to* 16°.		(Exp. No. 12) *Thermometer No.* 2 Surrounded by air. *Heated to* 80°, *and suspended in the open air warm to* 16°.	
Time elapsed	Heat lost	Time elapsed	Heat lost
	80°		80°
M. S.	0	M. S.	0
Not observed	70	Not observed	70
1 24	60	0 51	60
1 44	50	1 5	50
2 28	40	1 34	40
4 16	30	2 41	30
10 12 = total time employed in cooling from 70° to 30°.		6 11 = total time employed in cooling from 70° to 30°.	

the air (which appeared to be perfectly at rest, the windows and doors being all shut) was warm to the 16th degree of Reaumur's thermometer, and the times of cooling were observed in Table [5].

Here the difference in the conducting powers of air and of the Torricellian vacuum appears to be nearly the same as in the foregoing experiments, being as $6\frac{11}{60}$ to $10\frac{12}{60}$ inversely, or as 1000 to 605. I could not observe the time of cooling from 80° to 70°, being at that time busied in suspending the instruments.

As it might possibly be objected to the conclusions drawn from these experiments that, notwithstanding all the care that was taken in the construction of the two instruments made use of that they should be perfectly alike, yet they might in reality be so far different, either in shape or size, as to occasion a very sensible error in

TABLE 6

(Exp. No. 13) *Thermometer No.* 1		(Exp. No. 14) *The same Thermometer* (*No.* 1)	
Its bulb half an inch in diameter, shut up in the center of a glass globe, $1\frac{1}{2}$ inch in diameter, *void of air*, and hermetically sealed.		The glass globe, containing the bulb of the thermometer, being now *filled with air*, and hermetically sealed.	
Taken out of freezing water and plunged into boiling water.		*Taken out of freezing water and plunged into boiling water.*	
Time elapsed	Heat acquired	Time elapsed	Heat acquired
	0°		0°
M. S.	0	M. S.	0
0 55	10	0 32	10
0 55	20	0 32	20
1 7	30	0 43	30
1 15	40	0 50	40
1 29	50	1 1	50
2 2	60	1 24	60
3 21	70	2 38	70
13 44	80	10 25	80
24 48 = total time of heating from 0° to 80°.		18 5 = total time of heating from 0° to 80°.	
Total time from 0° to 70° = 11′ 4″.		Total time from 0° to 70° = 7′ 40″.	

the result of the experiments; to remove these doubts I made the following experiments:—

In the morning towards eleven o'clock, the weather being remarkably fine, the mercury in the barometer standing at 27 inches 11 lines, Reaumur's thermometer at 15°, and the hygrometer at 47°, I repeated the experiment No. 3 (of heating the thermometer No. 1 in boiling water, &c.), and immediately afterwards opened the cylinder containing the thermometer at its upper end, where it had been sealed, and letting the air into it, I resealed it hermetically, and repeated the experiment again with the same instrument, the thermometer being now surrounded with air, like the thermometer No. 2.

The result of these experiments, which may be seen in Table [6], shews evidently that the error arising from the difference of the shapes or dimensions of the two instruments in question was inconsiderable, if not totally imperceptible.

It appears, therefore, from these experiments, that the conducting power of common atmospheric air is to that of the Torricellian vacuum as $7\frac{40}{60}$ to $11\frac{4}{60}$ inversely, or as 1000 to 602; which differs but very little from the result of all the foregoing experiments.

Notwithstanding that it appeared, from the result of these last experiments, that any difference there might possibly have been in the forms or dimensions of the instruments No. 1 and No. 2 could hardly have produced any sensible error in the result of the experiments in question; I was willing, however, to see how far any considerable alterations of size in the instrument would affect the experiment: I therefore provided myself with another instrument, which I shall call *Thermometer No.* 3, different from those already described in size, and a little different in its construction.

The bulb of the thermometer was of the same form and size as in the instruments No. 1 and No. 2, that is to say, it was globular, and half an inch in diameter; but the glass globe, in the center of which it was confined, was much larger, being 3 inches $7\frac{1}{2}$ lines in diameter; and the bore of the tube of the thermometer was much finer, and consequently its length and the divisions of its scale were greater. The divisions were marked upon the tube with threads of silk of different colours at every tenth degree, from 0° to 80°, as in the before-mentioned instruments. The tube or cylinder belonging to the glass globe was 8 lines in diameter, a little longer than the tube of the thermometer, and perfectly cylindrical from its upper end to its junction with the globe, being without any choak; the thermometer being confined in the center of the globe by a different contrivance, which was as follows. To the opening of the cylinder was fitted a stopple of dry wood, covered with a coating of hard varnish, through the centre or axis of which passed the end of the tube of the thermometer; this stopple confined the tube in the axis of the cylinder at its upper end. To confine it at its lower end, there was fitted to it a small steel spring, a little below the point 0°; which, being fastened to the tube of the thermometer, had three elastic points projecting outwards, which, pressing against the inside of the cylinder, confined the thermometer in its place. The total length of this instrument, from the bottom of the globe to the upper end of the cylinder, was 18 inches, and the freezing point upon the thermometer fell about 3 inches above the bulb; consequently this point lay about $1\frac{1}{2}$ inch above the junction of the cylinder with the globe, when the thermometer was confined in its place, the center of its bulb coinciding with the center of the globe. Through the stopple which closed the end of the cylinder passed two small glass tubes, about a line in diameter, which being about a line longer than the stopple were closed occasionally with small stopples fitted to their bores. These tubes (which were fitted exactly in the holes bored in the great stopple of the cylinder to receive them, and fixed in their places with cement) served to convey air, or any other fluid, into the glass ball, without its being necessary to remove the stopple closing the end of the cylinder; which stopple, in order to prevent the position of the thermometer from being easily deranged, was cemented in its place.

I have been the more particular in the description of these instruments, as I conceive it to be absolutely necessary to have a perfect idea of them in order to judge of the experiments made with them, and of their results.

With the instrument last described (which I have called *Thermo-*

TABLE 7

(Exp. No. 15)

Thermometer No. 3

Its bulb half an inch in diameter, shut up in the center of a glass tube, 3 inches $7\frac{1}{2}$ lines in diameter, and surrounded by *air*.

Taken out of freezing water and plunged into boiling water.

Time elapsed		Heat acquired
		0°
M.	S.	0
0	33	10
0	38	20
0	54	30
0	51	40
1	7	50
1	28	60
2	28	70
9	0	80
16	59	= total time of heating from 0° to 80°

Time from 0° to 70° = 7′ 59″

(Exp. No. 5 and No. 6)

Thermometer No. 2

Its bulb half an inch in diameter, shut up in the center of a glass globe, $1\frac{1}{2}$ inch in diameter, and surrounded by *air*.

Taken out of freezing water and plunged into boiling water.

Time elapsed						Heat acquired
Exp. No. 5		Exp. No. 6		Medium		
						0°
M.	S.	M.	S.	M.	S.	0
0	30	0	30	0	30	10
0	35	0	37	0	36	20
0	41	0	41	0	41	30
0	49	0	53	0	51	40
1	1	0	59	1	0	50
1	24	1	20	1	22	60
2	45	2	25	2	35	70
9	10	9	38	9	24	80
16	55	17	3	16	59	= total time of heating from 0° to 80°

Time from 0° to 70° = 7′ 35″

meter No. 3) I made the following experiment. It was upon the 18th of July, 1785, in the afternoon, the weather variable, alternate clouds and sunshine; wind strong at S.E. with now and then a sprinkling of rain; barometer at 27 inches 10½ lines, thermometer at 18¼°, and hygrometer variable from 44° to extreme moisture.

In order to compare the result of the experiment made with this instrument with those made with the thermometer No. 2, I have placed together in the same table [Table 7] the different experiments made with them.

If the agreement of these experiments with the thermometers No. 2 and No. 3 surprised me, I was not less surprised with their disagreement in the experiment which follows:—

Experiment No. 16

Taking the thermometer No. 3 out of the boiling water, I immediately suspended it in the middle of a large room, where the air, which was quiet, was at the temperature of 18¼°R. and observed the times of cooling as follows:—

Time elapsed M.	S.	Heat lost
		80°
1	55	0
		70
0	12	60
0	33	50
2	15	40
4	0	30
9	55	= total time of cooling from 80° to 30°.

Time from 70° to 30° = 8′ 0″; but in the experiment No. 12, with the thermometer No. 2, the time employed in cooling from 70° to 30° was only 6′ 11″. In this experiment, with the thermometer No. 3, the time employed in cooling from 60° to 30° was 7′ 48″; but in the above-mentioned experiment, with the thermometer No. 2 it was only 5′ 20″. It is true, the air of the room was somewhat cooler when the former experiment was made, than when this latter was made, with the thermometer No. 3; but this difference of temperature, which was only $2\frac{1}{4}$° (in the former case the thermometer in the room standing at 16°, and in the latter at $18\frac{1}{4}$°), certainly could not have occasioned the whole of the apparent difference in the results of the experiments.

Does air receive Heat more readily than it parts with it? This is a question highly deserving of further investigation, and I hope to be able to give it a full examination in the course of my projected inquiries; but leaving it for the present, I shall proceed to give an account of the experiments which I have already made. Conceiving it to be a step of considerable importance towards coming at a further knowledge of the nature of Heat, to ascertain, by indisputable evidence, its passage through the Torricellian vacuum, and to determine, with as much precision as possible, the law of its motions in that medium; and being apprehensive that doubts might arise with respect to the experiments before described, on account of the contact of the tubes of the inclosed thermometers in the instruments made use of with the containing glass globes, or rather with their cylinders: by means of which (it might be suspected) that a certain quantity, if not all the Heat acquired, might possibly be communicated; to put this matter beyond all doubt, I made the following experiment.

In the middle of a glass body, of a pear-like form, about 8 inches long, and $2\frac{1}{2}$ inches in its greatest diameter, I suspended a small mercurial thermometer, $5\frac{1}{2}$ inches long, by a fine thread of silk, in such a manner that neither the bulb of the thermometer, nor its tube, touched the containing glass body in any part. The tube of the thermometer was graduated, and marked with fine threads of silk of different colours, bound round it, as in the thermometers belonging to the other instruments already described, and the thermometer was suspended in its place by means of a small steel spring, to which the end of the thread of silk which held the thermometer being attached, it (the spring) was forced into a small globular protuberance or cavity, blown in the upper extremity of the glass body, about half an inch in diameter, where, the spring remaining, the thermometer necessarily remained suspended in the axis of the glass body. There was an opening at the bottom of the glass body, through which the thermometer was introduced; and a barometrical tube being soldered to this opening, the inside of the glass body was voided of air by means of mercury; and this opening being afterwards sealed hermetically, and the barometrical tube being taken away, the thermometer was left suspended in a Torricellian vacuum.

In this instrument, as the inclosed thermometer did not touch the containing glass body in any part, on the contrary, being distant from its internal surface an inch or more in every part, it is clear that whatever Heat passed *into* or *out of* the thermometer must have passed *through* the surrounding Torricellian vacuum; for it cannot be supposed that the fine thread of silk, by which the thermometer was suspended, was capable of conducting any Heat at all, or at least any sensible quantity. I therefore flattered myself with hopes of being able, with the assistance of this instrument, to determine positively with regard to the passage of Heat in the Torricellian vacuum: and this I think I have done, notwithstanding an unfortunate accident that put it out of my power to pursue the experiment so far as I intended.

This instrument being fitted to a small stand or foot of wood, in such a manner that the glass body remained in a perpendicular situation, I placed it in my room, by the side of another inclosed thermometer (No. 2) which was surrounded by air, and observed the effects produced on it by the variation of Heat in the atmosphere. I soon discovered, by the motion of the mercury in the inclosed thermometer, that the Heat passed through the Torricellian vacuum; but it appeared plainly, from the sluggishness or great insensibility of the thermometer, that the Heat passed with much greater difficulty in this medium than in common air. I now plunged both the thermometers into a bucket of cold water; and I observed that the mercury in the thermometer surrounded by air descended much faster than that in the thermometer surrounded by the Torricellian vacuum. I took them out of the cold water, and plunged them into a vessel of hot water (having no conveniences at hand to repeat the experiment in due form with the freezing and with the boiling water); and the thermometer surrounded by the Torricellian vacuum appeared still to be much more insensible or sluggish than that surrounded by air.

These trials were quite sufficient to convince me of the passage of Heat in the Torricellian vacuum, and also of the greater difficulty of its passage in that medium than in common air; but not satisfied to rest my inquiries here, I took the first opportunity that offered, and set myself to repeat the experiments which I had before made with the instruments No. 1 and No. 2. I plunged this instrument into a mixture of pounded ice and water, where I let it remain till the mercury in the inclosed thermometer had descended to 0°; when, taking it out of this cold mixture, I plunged it suddenly into a vessel of boiling water, and prepared myself to observe the ascent of the mercury in the inclosed thermometer, as in the foregoing experiments; but unfortunately the moment the end of the glass body touched the boiling water, it cracked with the Heat at the point where it had been hermetically sealed, and the water rushing into the body spoiled the experiment: and I have not since had an opportunity of providing myself with another instrument to repeat it.

It having been my intention from the beginning to examine the conducting powers of the artificial airs or gases, the thermometer No. 3 was constructed with a view to those experiments; and having now provided myself with a stock of those different kinds of airs, I began with *fixed air*, with which, by means of water, I filled the globe and cylinder containing the thermometer; and stopping up the two holes in the great stopple closing the end of the cylinder, I exposed the instrument in freezing water till the mercury in the inclosed thermometer had descended to 0°; when, taking it out of the freezing water, I plunged it into a large vessel of boiling water, and prepared myself to observe the times of heating, as in the former cases; but an accident happened, which suddenly put a stop to the experiment. Immediately upon plunging the instrument into the boiling water, the mercury began to rise in the thermometer with such uncommon celerity that it had passed the first division upon the tube (which marked the 10th degree, according to Reaumur's scale) before I was aware of its being yet in motion; and having thus missed the opportunity of observing the time elapsed when the mercury arrived at that point, I was preparing to observe its passage of the next, when all of a sudden the stopple closing the end of the cylinder was blown up the chimney with a great explosion, and the thermometer, which, being cemented to it by its tube, was taken along with it, was broken to pieces, and destroyed in its fall.

This unfortunate experiment, though it put a stop for the time to the inquiries proposed, opened the way to other researches not less interesting. Suspecting that the explosion was occasioned by the rarefaction of the water which remained attached to the inside of the globe and cylinder after the operation of filling them with fixed air, and thinking it more than probable that the uncommon celerity with which the mercury rose in the thermometer was principally owing to the same cause, I was led to examine the conducting power of *moist air*, or air saturated with water.

For this experiment I provided myself with a new thermometer No. 4, the bulb of which, being of the same form as those already described (*viz.* globular), was also of the same size, or half an inch in diameter. To receive this thermometer a glass cylinder was provided, 8 lines in diameter, and about 14 inches long, and terminated at one end by a globe 1½ inch in diameter. In the center of this globe the bulb of the thermometer was confined, by means of the stopple which closed the end of the cylinder; which stopple, being near 2 inches long, received the end of the tube of the thermometer into a hole bored through its center or axis, and confined the thermometer in its place, without the assistance of any other apparatus. Through this stopple two other small holes were bored, and lined with thin glass tubes, as in the thermometer No. 3, opening a passage into the cylinder, which holes were occasionally stopped up with stopples of cork; but to prevent accidents, such as I have before experienced from an explosion, great care was taken not to press these stopples into their places with any considerable force, that they might the more easily be blown out by any considerable effort of the confined air, or vapour.

Though in this instrument the thermometer was not altogether so steady in its place as in the thermometers No. 1, No. 2, and No. 3, the elasticity of the tube, and the weight of the mercury in the bulb of the thermometer, occasioning a small vibration or trembling of the thermometer upon any sudden motion or jar; yet I preferred this method to the others, on account of the lower part of this thermometer being entirely free, or suspended in such a manner as not to touch, or have any communication with, the lower part of the cylinder of the globe; for though the quantity of Heat received by the tube of the thermometer at its contact with the cylinder at its choaks, in the instruments No. 1 and No. 2, or with the branches of the steel spring in No. 3, and from thence communicated to the bulb, must have been exceedingly small; yet I was desirous to prevent even that, and every other possible cause of error or inaccuracy.

TABLE 8

(Exp. No. 17) *Thermometer No.* 4 Surrounded by air dry to the 44th degree of the quill hygrometer of the Manheim Academy. *Taken out of freezing water and plunged into boiling water.*		(Exp. No. 18) *The same Thermometer (No.* 4) Surrounded by air rendered as moist as possible by wetting the inside of the cylinder and globe with water. *Taken out of freezing water and plunged into boiling water.*	
Time elapsed	Heat acquired 80°	Time elapsed	Heat acquired 80°
M. S.	0	M. S.	0
0 34	10	0 6	10
0 39	20	0 4	20
0 44	30	0 5	30
0 51	40	0 9	40
1 6	50	0 18	50
1 35	60	0 26	60
2 40	70	0 43	70
Not observed	80	7 45	80
8 9 = total time of heating from 0° to 70°.		1 51 = total time of heating from 0° to 70°.	

Does humidity augment the conducting power of air?

To determine this question I made the following experiments [Table 8], the weather being clear and fine, the mercury in the barometer standing at 27 inches 8 lines, the thermometer at 19°, and the hygrometer at 44°.

From these experiments it appears that the conducting power of air is very much increased by humidity. To see if the same result would obtain when the experiment was reversed, I now took the thermometer with the *moist air* out of the boiling water, and plunged it into freezing water; and moving it about continually from place to place in the freezing water, I observed the times of cooling, as set down in the following table. N.B. To compare the result of this experiment with those made with *dry air*, I have placed on one side in the following table [Table 9] the experiment in question, and on the other side the experiment No. 10, made with the thermometer No. 2.

Though the difference of the whole times of cooling from 80° to 10° in these two experiments appears to have been very small, yet

TABLE 9

(Exp. No. 19) *Thermometer No.* 4 Surrounded by moist air. *Taken out of boiling water and plunged into freezing water.*			(Exp. No. 10) *Thermometer No.* 2 Surrounded by dry air. *Taken out of boiling water and plunged into freezing water.*		
Time elapsed		Heat lost	Time elapsed		Heat lost
		80°			80°
M.	S.	0	M.	S.	0
0	4	70	0	33	70
0	14	60	0	34	60
0	31	50	0	44	50
0	52	40	0	55	40
1	22	30	1	18	30
2	3	20	1	57	20
4	2	10	3	40	10
9	8	= total time of cooling from 80° to 10°.	9	41	= total time of cooling from 80° to 10°.

the difference of the times taken up by the first twenty or thirty degrees from the boiling point is very remarkable, and shows with how much greater facility Heat passes in moist air than in dry air. Even the slowness with which the mercury in the thermometer No. 4 descended in this experiment from the 30th to the 20th, and from the 20th to the 10th degree, I attribute in some measure to the great conducting power of the moist air with which it was surrounded; for the cylinder containing the thermometer and the moist air being not wholly submerged in the freezing water, that part of it which remained out of the water was necessarily surrounded by the air of the atmosphere; which, being much warmer than the water, communicated of its Heat to the glass; which, passing from thence into the contained moist air as soon as that air became colder than the external air, was, through that medium, communicated to the bulb of the inclosed thermometer, which prevented its cooling so fast as it would otherwise have done. But when the weather becomes cold, I propose to repeat this experiment with variations, in such a manner as to put the matter beyond all doubt. In the mean time I cannot help observing, with what infinite wisdom and goodness Divine Providence appears to have guarded us against the evil effects of excessive Heat and Cold in the atmosphere; for if it were possible for the air to be equally damp during the severe cold of the winter months as it sometimes is in summer, its conducting power, and consequently its apparent coldness, when applied to our bodies, would be so much increased, by such an additional degree of moisture, that it would become quite intolerable; but, happily for us, its power to hold water in solution is diminished, and with it its power to rob us of our animal heat, in proportion as its coldness is increased. Everybody knows how very disagreeable a very moderate degree of cold is when the air is very damp; and from hence it appears, why the thermometer is not always a just measure of the apparent or sensible Heat of the atmosphere. If colds or catarrhs are occasioned by our bodies being robbed of our animal heat, the reason is plain why those disorders prevail most during the cold autumnal rains, and upon the breaking up of the frost in the spring. It is likewise plain from whence it is that sleeping in damp beds, and inhabiting damp houses, is so very dangerous; and why the evening air is so pernicious in summer and in autumn, and why it is not so during the hard frosts of winter. It has puzzled many very able philosophers and physicians to account for the manner in which the extraordinary degree or rather *quantity* of Heat is generated which an animal body is supposed to lose, when exposed to the cold of winter, above what it communicates to the surrounding atmosphere in warm summer weather; but is it not more than probable that the difference of the quantities of Heat, actually lost or communicated, is infinitely less than what they have imagined? These inquiries are certainly very interesting; and they are undoubtedly within the reach of well-contrived and well-conducted experiments. But taking my leave for the present of this curious subject of investigation, I hasten to the sequel of my experiments.

Finding so great a difference in the conducting powers of common air and of the Torricellian vacuum, I was led to examine the conducting powers of common air of different degrees of density. For this experiment I prepared the thermometer No. 4, by stopping up one of the small glass tubes passing through the stopple, and opening a passage into the cylinder, and by fitting a valve to the external overture of the other. The instrument, thus prepared, being put under the receiver of an air-pump, the air passed freely out of the globe and cylinder upon working the machine, but the valve above described prevented its return upon letting air into the receiver. The gage of the air-pump shewed the degree of rarity of the air under the receiver, and consequently of that filling the globe and cylinder, and immediately surrounding the thermometer.

With this instrument, the weather being clear and fine, the mercury in the barometer standing at 27 inches 9 lines, the thermometer at 15°, and the hygrometer at 47°, I made the following experiments.

The result of these experiments, I confess, surprised me not a little; but the discovery of truth being the sole object of my inquiries (having no favourite theory to defend) it brings no disappointment along with it, under whatever unexpected shape it

TABLE 10

(Exp. No. 20) *Thermometer No.* 4 Surrounded by common air, barometer standing at 27 inches 9 lines. *Taken out of freezing water, and plunged into boiling water.*		(Exp. No. 21) *Thermometer No.* 4 Surrounded by air rarefied by pumping till the barometer-gage stood at 6 inches $11\frac{1}{2}$ lines *Taken out of freezing water, and plunged into boiling water.*		(Exp. No. 22) *Thermometer No.* 4 Surrounded by air rarefied by pumping till the barometer-gage stood at 1 inch 2 lines *Taken out of freezing water, and plunged into boiling water.*	
Time elapsed	Heat acquired	Time elapsed	Heat acquired	Time elapsed	Heat acquired
	0°		0°		0°
M. S.		M. S.		M. S.	
0 31	10	0 31	10	0 29	10
0 40	20	0 38	20	0 36	20
0 41	30	0 44	30	0 49	30
0 47	40	0 51	40	1 1	40
1 4	50	1 7	50	1 1	50
1 25	60	1 19	60	1 24	60
2 28	70	2 27	70	2 31	70
10 17	80	10 21	80	Not observed	80
7 36 = total time of heating from 0° to 70°		7 37 = total time of heating from 0° to 70°		7 51 = total time of heating from 0° to 70°	

may appear. I hope that further experiments may lead to the discovery of the cause why there is so little difference in the conducting powers of air of such very different degrees of rarity, while there is so great a difference in the conducting powers of air, and of the Torricellian vacuum. At present I shall not venture any conjectures upon the subject; but in the mean time I dare to assert that the experiments I have made may be depended on.

The time of my stay at Manheim being expired (having had the honour to attend thither his most Serene Highness the Elector Palatine, reigning Duke of Bavaria, in his late journey) I was prevented from pursuing these inquiries further at that time; but I shall not fail to recommence them the first leisure moment I can find, which I fancy will be about the beginning of the month of November. In the mean time, to enable myself to pursue them with effect, I am sparing neither labor nor expence to provide a complete apparatus necessary for my purpose; and his Electoral Highness has been graciously pleased to order M. ARTARIA (who is in his service) to come to Munich to assist me. With such a patron as his most Serene Highness, and with such an assistant as ARTARIA, I shall go on in my pursuits with cheerfulness. Would to God that my labours might be as useful to others as they will be pleasant to me!

I shall conclude this chapter with a short account of some experiments I have made to determine the conducting powers of water and of mercury; and with a table, showing at one view the conducting powers of all the different mediums which I have examined.

Having filled the glass globe inclosing the bulb of the thermometer No. 4, first with water, and then with mercury, I made the following experiments [Table 11], to ascertain the conducting powers of those two fluids.

The total times of heating from 0° to 70° in the three experiments with mercury being 41 seconds, 31 seconds, and 48 seconds, the mean of these times is 40 seconds; and as in the experiment with water the time employed in acquiring the same degree of Heat was 1′ 57″ = 117 seconds, it appears from these experiments that the conducting power of mercury to that of water, under the

TABLE 11

(Exp. No. 23) *Thermometer No.* 4 Surrounded by water *Taken out of freezing water, and plunged into boiling water.*		(Exp. No. 24, 25 and 26) *Thermometer No.* 4 Surrounded by mercury *Taken out of freezing water, and plunged into boiling water.*			
		Time elapsed			Heat acquired
Time elapsed	Heat acquired	Exp. No. 24	Exp. No. 25	Exp. No. 26	
	0°				0°
M. S.	0	M. S.	M. S.	M. S.	0
0 19	10	0 5	0 5	0 5	10
0 8	20	0 4	0 2	0 5	20
0 9	30	0 2	0 2	0 4	30
0 11	40	0 4	0 5	0 5	40
0 15	50	0 4	0 4	0 7	50
0 21	60	0 7	0 4	0 8	60
0 34	70	0 15	0 9	0 14	70
2 13	80	Not observed	0 58	Not observed	80
1 57 = total time of heating from 0° to 70°		0 41	0 31	0 48 = total times of heating from 0° to 70°.	

circumstances described, is as $36\frac{2}{3}$ to 117 inversely, or as 1000 to 342. And hence it is plain, why mercury *appears* so much hotter, and so much colder, to the touch than water, when in fact it is of the same temperature: for the force or violence of the sensation of what appears *hot* or *cold* depends not entirely upon the temperature of the body exciting in us those sensations, or upon the degree of Heat it actually possesses, but upon the *quantity* of Heat it is capable of communicating to us, or receiving from us in any given short period of time, or as the intensity of the communication; and this depends in a great measure upon the conducting powers of the bodies in question.

The sensation excited in us when we touch anything that appears to us to be *hot* is the entrance of heat into our bodies; that of *cold* is its exit; and whatever contributes to facilitate or accelerate this communication adds to the violence of the sensation. And this is another proof that the thermometer cannot be a just measure of the intensity of the *sensible* Heat, or Cold, existing in bodies; or rather, that the touch does not afford us a just indication of their *real* temperatures.

In determining the relative conducting powers of these mediums, I have compared the times of the heating of the thermometer from 0° to 70° instead of taking the whole times from 0° to 80°, and this I have done on account of the small variation in the Heat of the boiling water arising from the variation of the weight of the atmosphere, and also on account of the very slow motion of the mercury between the 70th and the 80th degrees, and the difficulty of determining the precise moment when the mercury arrives at the 80th degree.

Taking now the conducting power of mercury = 1000, the conducting powers of the other mediums, as determined by these experiments, will be as follows, *viz.*:—

Mercury	1000
Moist air	330
Water	313
Common air, density = 1	$80\frac{41}{100}$
Rarefied air, density = $\frac{1}{4}$	$80\frac{23}{100}$
Rarefied air, density = $\frac{1}{24}$	78
The Torricellian vacuum	65

TABLE 12

Therm. No. 1	*Thermometer No.* 4						
Taken out of freezing water and plunged into boiling water.							
Time elapsed							
Torricellian vacuum (Exp. No. 3, 4, and 13)	Common air, density = 1 (Exp. No. 20)	Rarefied air, density = $\frac{1}{4}$ (Exp. No. 21)	Rarefied air, density = $\frac{1}{24}$ (Exp. No. 22)	Moist air (Exp. No. 18)	Water (Exp. No. 23)	Mercury (Exp. No. 24, 25, and 26)	Heat acquired
M. S.	M. S.	M. S.	M. S.	M. S.	M. S.	M. S.	0° 0
0 52	0 31	0 31	0 29	0 6	0 19	0 5	10
0 58	0 40	0 38	0 36	0 4	0 8	0 $3\frac{2}{3}$	20
1 3	0 41	0 44	0 49	0 5	0 9	0 $2\frac{2}{3}$	30
1 18	0 47	0 51	1 1	0 9	0 11	0 $4\frac{2}{3}$	40
1 25	1 4	1 7	1 1	0 18	0 15	0 5	50
1 58	1 25	1 19	1 24	0 26	0 21	0 $6\frac{1}{3}$	60
3 19	2 28	2 27	2 31	0 43	0 34	0 $12\frac{2}{3}$	70
11 57	10 17	10 21	——	7 45	2 13	0 58	80
10 53	7 36	7 37	7 51	1 51	1 57	0 40	= total

times of heating from 0° to 70°

And in these proportions are the quantities of Heat which these different mediums are capable of transmitting in any given time; and consequently these numbers express the relative *sensible* temperatures of the mediums, as well as their conducting powers. How far these decisions will hold good under a variation of circumstances, experiment only can determine. This is certainly a subject of investigation not less curious in itself than it is interesting to mankind; and I wish that what I have done may induce others to turn their attention to this long neglected field of experimental inquiry. For my own part I am determined not to quit it.

In the further prosecution of these inquiries, I do not mean to confine myself solely to the determining of the conducting powers of Fluids; on the contrary, Solids, and particularly such bodies as are made use of for cloathing, will be principal subjects of my future experiments. I have indeed already begun these researches, and have made some progress in them; but I forbear to anticipate a matter which will be the subject of a future communication.

The confining and directing of Heat are objects of such vast importance in the economy of human life, that I have been induced to confine my researches chiefly to those points, conceiving that very great advantages to mankind could not fail to be derived from the discovery of any new facts relative to these operations.

If the laws of the communication of Heat from one body to another were known, measures might be taken with certainty, in all cases, for confining it, and directing its operations, and this would not only be productive of great economy in the articles of fuel and cloathing, but would likewise greatly increase the comforts and conveniences of life,—objects of which the philosopher should never lose sight.

The route which I have followed in this inquiry is that which I thought bid fairest to lead to useful discoveries. Without embarrassing myself with any particular theory, I have formed to myself a plan of experimental investigation, which I conceived would conduct me to the knowledge of *certain facts*, of which we are now ignorant, or very imperfectly informed, and with which it is of consequence that we should be made acquainted.

The first great object which I had in view in this inquiry was to ascertain, if possible, the cause of the warmth of certain bodies, or the circumstances upon which their power of confining Heat depends. This, in other words, is no other than to determine the cause of the conducting and non-conducting power of bodies, with regard to Heat.

To this end, I began by determining by actual experiment the relative conducting powers of various bodies of very different natures, both Fluids and Solids; of some of which experiments I have already given an account in the paper above mentioned, which is published in the Transactions of the Royal Society for the year 1786: I shall now, taking up the matter where I left it, give the continuation of the history of my researches.

Having discovered that the Torricellian vacuum is a much worse conductor of Heat than common air, and having ascertained the relative conducting powers of air, of water, and of mercury, under different circumstances, I proceeded to examine the conducting powers of various *solid bodies*, and particularly of such substances as are commonly made use of for cloathing.

The method of making these experiments was as follows: a mercurial thermometer (see Fig. 4 [Plate 9]), whose bulb was about $\frac{55}{100}$ of an inch in diameter, and its tube about 10 inches in length, was suspended in the axis of a cylindrical glass tube, about $\frac{3}{4}$ of an inch in diameter, ending with a globe $1\frac{6}{10}$ inch in diameter, in such a manner that the center of the bulb of the thermometer occupied the center of the globe; and the space between the internal surface of the globe and the surface of the bulb of the thermometer being filled with the substance whose conducting power was to be determined, the instrument was heated in boiling water, and afterwards, being plunged into a freezing mixture of pounded ice and water, the times of cooling were observed, and noted down.

The tube of the thermometer was divided at every tenth degree from 0°, or the point of freezing, to 80°, that of boiling water; and these divisions being marked upon the tube with the point of a diamond, and the cylindrical tube being left empty, the height of the mercury in the tube of the thermometer was seen through it.

The thermometer was confined in its place by means of a stopple of cork, about 1½ inch long, fitted to the mouth of the cylindrical tube, through the center of which stopple the end of the tube of the thermometer passed, and in which it was cemented.

The operation of introducing into the globe the substances whose conducting powers are to be determined, is performed in the following manner: the thermometer being taken out of the cylindrical tube, about two thirds of the substance which is to be the subject of the experiment are introduced into the globe; after which, the bulb of the thermometer is introduced a few inches into the cylinder; and, after it, the remainder of the substance being placed round about the tube of the thermometer; and, lastly, the thermometer being introduced farther into the tube, and being brought into its proper place, that part of the substance which, being introduced last, remains in the cylindrical tube above the bulb of the thermometer, is pushed down into the globe, and placed equally round the bulb of the thermometer by means of a brass wire which is passed through holes made for that purpose in the stopple closing the end of the cylindrical tube.

As this instrument is calculated merely for measuring the passage of Heat in the substance whose conducting power is examined, I shall give it the name of *passage-thermometer*, and I shall apply the same appellation to all other instruments constructed upon the same principles, and for the same use, which I may in future have occasion to mention; and as this instrument has been so particularly described, both here, and in my former paper upon the subject of Heat, in speaking of any others of the same kind in future it will not be necessary to enter into such minute details. I shall, therefore, only mention their *sizes*, or the diameters of their bulbs, the diameters of their globes, the diameters of their cylinders, and the lengths and divisions of their tubes, taking it for granted that this will be quite sufficient to give a clear idea of the instrument.

In most of my former experiments, in order to ascertain the conducting power of any body, the body being introduced into the globe of the passage-thermometer, the instrument was cooled to the temperature of freezing water, after which, being taken out of the ice-water, it was plunged suddenly into boiling water, and the times of heating from ten to ten degrees were observed and noted; and I said that these times were as the conducting power of the body inversely; but in the experiments of which I am now about to give an account, I have in general reversed the operation; that is to say, instead of observing the times of heating, I have first heated the body in boiling water, and then plunged it into a mixture of pounded ice and ice-cold water, I have noted the times taken up in cooling.

I have preferred this last method to the former, not only on account of the greater ease and convenience with which a thermometer, plunged into ice and water, may be observed, than when placed in a vessel of boiling water, and surrounded by hot steam, but also on account of the greater accuracy of the experiment, the heat of boiling water varying with the variations of the pressure of the atmosphere; consequently, the experiments made upon different days will have different results, and of course, strictly speaking, cannot be compared together; but the temperature of pounded ice and water is ever the same, and of course the results of the experiments are uniform.

In heating the thermometer, I did not in general bring it to the temperature of the boiling water, as this temperature, as I have just observed, is variable; but when the mercury had attained the 75° of its scale, I immediately took it out of the boiling water, and plunged it into the ice and water; or, which I take to be still more accurate, suffering the mercury to rise a degree or two above 75°, and then taking it out of the boiling water, I held it over the vessel containing the pounded ice and water, ready to plunge it into that mixture the moment the mercury, descending, passes the 75°.

Having a watch at my ear which beat half seconds (which I counted), I noted the time of the passage of the mercury over the divisions of the thermometer, marking 70° and every tenth degree from it, descending to 10° of the scale. I continued the cooling to

0°, or the temperature of the ice and water, in very few instances, as this took up much time, and was attended with no particular advantage, the determination of the times taken up in cooling 60 degrees of Reaumur's scale—that is to say, from 70° to 10°—being quite sufficient to ascertain the conducting power of any body whatever.

During the time of cooling in ice and water, the thermometer was constantly moved about in this mixture from one place to another; and there was always so much pounded ice mixed with the water that the ice appeared above the surface of the water,—the vessel, which was a large earthen jar, being first quite filled with pounded ice, and the water being afterwards poured upon it, and fresh quantities of pounded ice being added as the occasion required.

Having described the apparatus made use of in these experiments, and the manner of performing the different operations, I shall now proceed to give an account of the experiments themselves.

My first attempt was to discover the relative conducting powers of such substances as are commonly made use of for cloathing; accordingly, having procured a quantity of *raw silk*, as spun by the worm, *sheep's-wool, cotton-wool, linen* in the form of the finest lint, being the scrapings of very fine Irish linen, the finest part of the *fur of the beaver* separated from the skin, and from the long hair, the finest part of the *fur of a white Russian hare*, and *eider-down*,—I introduced successively 16 grains in weight of each of these substances into the globe of the passage-thermometer, and placing it carefully and equally round the bulb of the thermometer, I heated the thermometer in boiling water, as before described, and taking it out of the boiling water, plunged it into pounded ice and water, and observed the times of cooling.

But as the interstices of these bodies thus placed in the globe were filled with air, I first made the experiment with air alone, and took the result of that experiment as a standard by which to compare all the others; the results of three experiments with air were as in Table [13]:

TABLE 13

The Bulb of the Thermometer surrounded by Air				
Heat lost	Exp. No. 1	Exp. No. 2	Heat acquired	Exp. No. 3
	Time elapsed	Time elapsed		Time elapsed
70°	—	—	10°	—
60	38″	38″	20	39″
50	46	46	30	43
40	59	59	40	53
30	80	79	50	67
20	122	122	60	96
10	231	230	70	175
Total times	576	574	—	473

The following table [Table 14] shews the results of the experiments with the various substances therein mentioned:—

TABLE 14

Heat lost	Air	Raw silk, 16 grs.	Sheep's wool, 16 grs.	Cotton-wool, 16 grs.	Fine lint, 16 grs.	Beavers' fur, 16 grs.	Hares' fur, 16 grs.	Eider-down, 16 grs.
	Exp. 1	Exp. 4	Exp. 5	Exp. 6	Exp. 7	Exp. 8	Exp. 9	Exp. 10
70°	—	—	—	—	—	—	—	—
60	38″	94″	79″	83″	80″	99″	97″	98″
50	46	110	95	95	93	116	117	116
40	59	133	118	117	115	153	144	146
30	80	185	162	152	150	185	193	192
20	122	273	238	221	218	265	270	268
10	231	489	426	378	376	478	494	485
Total times	576	1284	1118	1046	1032	1296	1315	1305

Now the *warmth* of a body, or its power to confine Heat, being as its power of resisting the passage of Heat through it (which I shall call its *non-conducting power*); and the time taken up by any body in cooling, which is surrounded by any medium through which the Heat is obliged to pass, being, *caeteris paribus*, as the resistance which the medium opposes to the passage of the Heat, it appears that the *warmth* of the bodies mentioned in the foregoing table are as the times of cooling,—the *conducting powers* being inversely as those times, as I have formerly shown.

From the results of the foregoing experiments it appears that, of the seven different substances made use of, hares' fur and eider-down were the warmest; after these came beavers' fur, raw silk, sheep's-wool, cotton-wool, and, lastly, lint, or the scrapings of fine linen; but I acknowledge that the differences in the warmth of these substances were much less than I expected to have found them.

Suspecting that this might arise from the volumes or solid contents of the substances being different (though their weights were the same), arising from the difference of their specific gravities; and as it was not easy to determine the specific gravities of these substances with accuracy, in order to see how far any known difference in the volume or quantity of the same substance, confined always in the same space, would add to or diminish the time of cooling, or the apparent warmth of the covering, I made the three following experiments.

In the first, the bulb of the thermometer was surrounded by 16 grains of eider-down; in the second by 32 grains; and in the third by 64 grains; and in all these experiments the substance was made to occupy exactly the same space, *viz.* the whole internal capacity of the glass globe, in the center of which the bulb of the thermometer was placed; consequently, the thickness of the covering of the thermometer remained the same, while its density was varied in proportion to the numbers 1, 2, and 4.

The results of these experiments were as Table [15].

Without stopping at present to draw any particular conclusions from the results of these experiments, I shall proceed to give an

TABLE 15

The Bulb of the Thermometer being surrounded by Eider-down			
Heat lost	16 grains	32 grains	64 grains
	(Exp. No. 11)	(Exp. No. 12)	(Exp. No. 13)
70°	—	—	—
60	97″	111″	112″
50	117	128	130
40	145	157	165
30	192	207	224
20	267	304	326
10	486	565	658
Total times	1304	1472	1615

account of some others, which will afford us a little further insight into the nature of some of the circumstances upon which the warmth of covering depends.

Finding, by the last experiments, that the density of the covering added so considerably to the warmth of it, its thickness remaining the same, I was now desirous of discovering how far the internal structure of it contributed to render it more or less pervious to Heat, its thickness and quantity of matter remaining the same. By internal structure, I mean the disposition of the parts of the substance which forms the covering; thus they may be extremely divided, or very fine, as raw silk as spun by the worms, and they may be equally distributed through the whole space they occupy; or they may be coarser, or in larger masses, with larger interstices, as the ravelings of cloth, or cuttings of thread.

If Heat passed *through* the substances made use of for covering, and if the warmth of the covering depended solely upon the difficulty which the Heat meets with in its passage through the substances, *or solid parts*, of which they are composed,—in that case, the warmth of covering would be always, *caeteris paribus*, as the quantity of materials of which it is composed; but that this is not the case, the following, as well as the foregoing, experiments clearly evince.

Having, in the experiment No. 4, ascertained the warmth of 16 grains of raw silk, I now repeated the experiment with the same quantity, or weight, of the ravelings of white taffety, and afterwards with a like quantity of common sewing-silk, cut into lengths of about two inches.

Table [16] shows the results of these three experiments:

TABLE 16

Heat lost	Raw silk, 16 grs.	Ravelings of taffety, 16 grs.	Sewing-silk cut into lengths, 16 grs.
	Exp. 4	Exp. 14	Exp. 15
70°	—	—	—
60	94″	90″	67″
50	110	106	79
40	133	128	99
30	185	172	135
20	273	246	195
10	489	427	342
Total times	1284	1169	917

Here, notwithstanding that the quantities of the silk were the same in the three experiments, and though in each of them it was made to occupy the same space, yet the warmth of the coverings which were formed were very different, owing to the different disposition of the material.

The raw silk was very fine, and was very equally distributed through the space it occupied, and it formed a warm covering.

The ravelings of taffety were also fine, but not so fine as the raw silk, and of course the interstices between its threads were greater, and it was less warm; but the cuttings of sewing-silk were very coarse, and consequently it was very unequally distributed in the space in which it was confined; and it made a very bad covering for confining Heat.

It is clear from the results of the last five experiments, that the air which occupies the interstices of bodies, made use of for covering, acts a very important part in the operation of confining Heat; yet I shall postpone the examination of that circumstance till I shall have given an account of several other experiments, which, I think, will throw still more light upon that subject.

But, before I go any further, I will give an account of three experiments, which I made, or, rather, the same experiment which I repeated three times the same day, in order to see how far they may be depended on, as being regular in their results.

The glass globe of the passage-thermometer being filled with 16 grains of cotton-wool, the instrument was heated and cooled three times successively, when the times of cooling were observed as in Table [17]:—

TABLE 17

Heat lost	Exp. 16	Exp. 17	Exp. 18
70°	—	—	—
60	82″	84″	83″
50	96	95	95
40	118	117	116
30	152	153	151
20	221	221	220
10	380	377	377
Total times	1049	1047	1082

The differences of the times of cooling in these three experiments were extremely small; but regular as these experiments appear to have been in their results, they were not more so than the other experiments made in the same way, many of which were repeated two or three times, though, for the sake of brevity, I have put them down as single experiments.

But to proceed in the account of my investigations relative to the causes of the warmth of warm cloathing. Having found that the fineness and equal distribution of a body or substance made use of to form a covering to confine Heat contributes so much to the warmth of the covering, I was desirous, in the next place, to see the effect of condensing the covering, its quantity of matter remaining the same, but its thickness being diminished in proportion to the increase of its density.

The experiment I made for this purpose was as follows: I took 16 grains of common sewing-silk, neither very fine nor very coarse, and winding it about the bulb of the thermometer in such a manner that it entirely covered it, and was as nearly as possible of the same thickness in every part, I replaced the thermometer in its cylinder and globe, and heating it in boiling water, cooled it in ice and water, as in the foregoing experiments. The results of the experiment were as may be seen in Table [18]; and in order that it may be compared with those made with the same quantity of silk differently disposed of, I have placed those experiments by the side of it:—

TABLE 18

Heat lost	Raw silk, 16 grs.	Fine ravelings of taffety, 16 grs.	Sewing-silk cut into lengths, 16 grs.	Sewing-silk, 16 grs. wound round the bulb of the thermometer
	Exp. No. 4	Exp. No. 14	Exp. No. 15	Exp. No. 19
70°	—	—	—	—
60	94″	90″	67″	46″
50	110	106	79	62
40	133	128	99	85
30	185	172	135	121
20	273	246	195	191
10	489	427	342	399
Total times	1284	1169	917	904

It is not a little remarkable, that, though the covering formed a sewing-silk wound round the bulb of the thermometer in the 19th experiment appeared to have so little power of confining the Heat when the instrument was very hot, or when it was first plunged into the ice and water, yet afterwards, when the Heat of the thermometer approached much nearer to that of the surrounding medium, its power of confining the Heat which remained in the bulb of the thermometer appeared to be even greater than that of the silk in the experiment No. 15, the time of cooling from 20° to 10° being in the one 399″, and in the other 342″. The same appearance was observed in the following experiments, in which the bulb of the thermometer was surrounded by threads of *wool*, of *cotton*, and of *linen*, or *flax*, wound round it, in the like manner as the sewing-silk was wound round it in the last experiment.

The following table [Table 19] shows the results of these experiments, with the threads of various kinds; and that they may the more easily be compared with those made with the same quantity of the same substances in a different form, I have placed the accounts of these experiments by the side of each other. I have also added the account of an experiment, in which 16 grains of fine linen cloth were wrapped round the bulb of the thermometer,

TABLE 19

Heat lost	*Sheep's-wool*, 16 grains, surrounding the bulb of the thermometer	*Woollen thread*, 16 grains, wound round the bulb of the thermometer	*Cotton-wool*, 16 grains, surrounding the bulb of the thermometer	*Cotton thread*, 16 grains, wound round the bulb of the thermometer	*Lint*, 16 grains, surrounding the bulb of the thermometer	*Linen thread*, 16 grains, wound round the bulb of the thermometer	*Linen cloth*, 16 grains, wrapped round the bulb of the thermometer
	Exp. 5	Exp. 20	Exp. 6	Exp. 21	Exp. 7	Exp. 22	Exp. 23
70°	—	—	—	—	—	—	—
60	79″	46″	83″	45″	80″	46″	42″
50	95	63	95	60	93	62	56
40	118	89	117	83	115	83	74
30	162	126	152	115	150	117	108
20	238	200	221	179	218	180	168
10	426	410	378	370	376	385	338
Total time	1118	934	1046	852	1032	873	783

going round it nine times, and being bound together at the top and bottom of it, so as completely to cover it.

That thread wound tight round the bulb of the thermometer should form a covering less warm than the same quantity of wool, or other raw materials of which the thread is made, surrounding the bulb of the thermometer in a more loose manner, and consequently occupying a greater space, is no more than what I expected, from the idea I had formed of the causes of the warmth of covering; but I confess I was much surprised to find that there is so great a difference in the relative warmth of these two coverings, when they are employed to confine great degrees of Heat, and when the Heat they confine is much less in proportion to the temperature of the surrounding medium. This difference was very remarkable; in the experiments with sheep's wool, and with woollen thread, the warmth of the covering formed of 16 grains of the former was to that formed of 16 grains of the latter, when the bulb of the thermometer was heated to 70° and cooled to 60°, as 79 to 46 (the surrounding medium being at 0°); but afterwards, when the thermometer had only fallen from 20° to 10° of Heat, the warmth of the wool was to that of the woollen thread only as 426 to 410; and in the experiments with lint, and with linen thread, when the Heat was much abated, the covering of the thread appeared to be even warmer than that of the lint, though in the beginning of the experiments, when the Heat was much greater, the lint was warmer than the thread, in the proportion of 80 to 46.

From hence it should seem that a covering may, under certain circumstances, be very good for confining small degrees of warmth, which would be but very indifferent when made use of for confining a more intense Heat, and *vice versa*. This, I believe, is a new fact; and I think the knowledge of it may lead to further discoveries relative to the causes of the warmth of coverings, or the manner in which Heat makes its passage through them. But I forbear to enlarge upon this subject, till I shall have given an account of several other experiments, which I think throw more light upon it, and which will consequently render the investigation easier and more satisfactory.

With a view to determine how far the power which certain bodies appear to possess of confining Heat, when made use of as covering, depends upon the natures of those bodies, considered as chymical substances, or upon the chymical principles of which they are composed, I made the following experiments.

As charcoal is supposed to be composed almost entirely of phlogiston, I thought that, if that principle was the cause either of the conducting power or the non-conducting power of the bodies which contain it, I should discover it by making the experiment with charcoal, as I had done with various other bodies. Accordingly, having filled the globe of the passage-thermometer with 176 grains of that substance in very fine powder (it having been pounded in a mortar, and sifted through a fine sieve), the bulb of the thermometer being surrounded by this powder, the instrument was heated in boiling water, and being afterwards plunged into a mixture of pounded ice and water, the times of cooling were observed as mentioned in the following table [Table 20]. I afterwards repeated the experiment with lampblack, and with very pure and very dry wood-ashes; the results of which experiments were as in Table 20.

TABLE 20

	The Bulb of the Thermometer surrounded by			
Heat lost	176 grains of fine powder of charcoal	176 grains of fine powder of charcoal	195 grains of lampblack	307 grains of pure dry wood-ashes
	Exp. No. 24	Exp. No. 25	Exp. No. 26	Exp. No. 27
70°	—	—	—	—
60	79″	91″	124″	96″
50	95	91	118	92
40	100	109	134	107
30	139	133	164	136
20	196	192	237	185
10	331	321	394	311
Total times	940	937	1171	927

The experiment No. 25 was simply a repetition of that numbered 24, and was made immediately after it; but, in moving the thermometer about in the former experiment, the powder of charcoal which filled the globe was shaken a little together, and to this circumstance I attribute the difference in the results of the two experiments.

In the experiments with lampblack and with wood-ashes, the times taken up in cooling from 70° to 60° were greater than those employed in cooling from 60° to 50°; this most probably arose from the considerable quantity of Heat contained by these substances, which was first to be disposed of, before they could receive and communicate to the surrounding medium that which was contained by the bulb of the thermometer.

The next experiment I made was with *semen lycopodii*, commonly called witch-meal, a substance which possesses very extraordinary properties. It is almost impossible to wet it; a quantity of it strewed upon the surface of a basin of water, not only swims upon the water without being wet, but it prevents other bodies from being wet which are plunged into the water through it; so that a piece of money, or other solid body, may be taken from the bottom of the basin by the naked hand without wetting the hand; which is one of the tricks commonly shown by the jugglers in the country: this meal covers the hand, and, descending along with it to the bottom of the basin, defends it from the water. This substance has the appearance of an exceeding fine, light, and very moveable yellow powder, and it is very inflammable; so much so, that, being blown out of a quill into the flame of a candle, it flashes like gunpowder, and it is made use of in this manner in our theatres for imitating lightning.

Conceiving that there must have been a strong attraction between this substance and air, and suspecting, from some circumstances attending some of the foregoing experiments, that the warmth of a covering depends not merely upon the fineness of the substance of which the covering is formed, and the disposition of its parts, but that it arises in some measure from a certain attraction between the substance and the air which fills its interstices, I thought that an experiment with *semen lycopodii* might possibly throw some light upon this matter; and in this opinion I was not altogether mistaken, as will appear by the results of the three following experiments [Table 21].

TABLE 21

The Bulb of the Thermometer surrounded by 256 Grs. of *Semen Lycopodii*

Heat lost	Cooled	Cooled	Heat acquired	Heated
	Exp. No. 28	Exp. No. 29		Exp. No. 30
70°	—	—	0°	—
60	146″	157″	10	230″
50	162	162	20	68
40	175	170	30	63
30	209	203	40	76
20	284	288	50	121
10	502	513	60	316
—	—	—	70	1585
Total times	1478	1491	—	2459

In the last experiment (No. 30), the result of which was so very extraordinary, the instrument was cooled to 0° in thawing ice, after which it was plunged suddenly into boiling water, where it remained till the inclosed thermometer had acquired the Heat of 70°, which took up no less than 2459 seconds, or above 40 minutes; and it had remained in the boiling water full a minute and a half before the mercury in the thermometer showed the least sign of rising. Having at length been put into motion, it rose very rapidly 40 or 50 degrees, after which its motion gradually abating became so slow, that it took up 1585 seconds, or something more than 26 minutes, in rising from 60° to 70°, though the temperature of the medium in which it was placed during the whole of this time was very nearly 80°; the mercury in the barometer standing but little short of 27 Paris inches.

All the different substances which I had yet made use of in these experiments for surrounding or covering the bulb of the thermometer, fluids excepted, had, in a greater or in a less degree confined the Heat, or prevented its passing into or out of the thermometer so rapidly as it would have done, had there been nothing but air in the glass globe, in the center of which the bulb of the thermometer was suspended. But the great question is, how, or in what manner, they produced this effect?

And first, it was not in consequence of their own nonconducting powers, simply considered; for if, instead of being only bad conductors of Heat, we suppose them to have been totally impervious to Heat, their volumes or solid contents were so exceedingly small in proportion to the capacity of the globe in which they were placed, that, had they had no effect whatever upon the air filling their interstices, that air would have been sufficient to have conducted all the Heat communicated in less time than was actually taken up in the experiment.

The diameter of the globe being 1.6 inches, its contents amounted to 2.14466 cubic inches; and the contents of the bulb of the thermometer being only 0.08711 of a cubic inch (its diameter being 0.55 of an inch), the space between the bulb of the thermometer and the internal surface of the globe amounted to 2.14466 − 0.8711 = 2.05755 cubic inches; the whole of which space was occupied by the substances by which the bulb of the thermometer was surrounded in the experiments in question.

But though these substances occupied this space, they were far from *filling it*; by much the greater part of it being filled by the air which occupied the interstices of the substances in question. In the experiment No. 4, this space was occupied by 16 grains of raw silk; and as the specific gravity of raw silk is to that of water as 1734 to 1000, the volume of this silk was equal to the volume of 9.4422 grains of water; and as 1 cubic inch of water weighs 253.185 grains, its volume was equal to $\frac{9.4422}{253.1850} = 0.037294$ of a cubic inch; and, as the space it occupied amounted to 2.05755 cubic inches, it appears that the silk filled no more than about $\frac{1}{55}$ part of the space in which it was confined, the rest of that space being filled with air.

In the experiment No. 1, when the space between the bulb of the thermometer and the glass globe, in the center of which it was confined, was filled with nothing but air, the time taken up by the thermometer in cooling from 70° to 10° was 576 seconds; but in the experiment No. 4, when this same space was filled with 54 parts air, and 1 part raw silk, the time of cooling was 1284 seconds.

Now, supposing that the silk had been totally incapable of conducting any Heat at all, if we suppose, at the same time, that it had no power to prevent the air remaining in the globe from conducting it, in that case its presence in the globe could only have prolonged the time of cooling in proportion to the quantity of the air it had displaced to the quantity remaining, that is to say, as 1 is to 54, or a little more than 10 seconds. But the time of cooling was actually prolonged 708 seconds (for in the experiment No. 1 it was 576 seconds, and in the experiment No. 4 it was 1284 seconds, as has just been observed); and this shows that the silk not only did not conduct the Heat itself, but that it prevented the air by which its interstices were filled from conducting it; or, at least, it greatly weakened its power of conducting it.

The next question which arises is, how air can be prevented from conducting Heat? and this necessarily involves another, which is, How does air conduct Heat?

If air conducted Heat, as it is probable that the metals and water, and all other solid bodies and unelastic fluids, conduct it,—that is to say, if, its particles remaining in their places, the Heat passed from one particle to another, through the whole mass, as there is no reason to suppose that the propagation of Heat is necessarily in right lines, I cannot conceive how the interposition of so small a quantity of any solid body as $\frac{1}{55}$ part of the volume of the air could have effected so remarkable a diminution of the conducting power of the air, as appeared in the experiment (No. 4) with raw silk, above-mentioned.

If air and water conducted Heat in the same *manner*, it is more than probable that their conducting powers might be impaired by the same means; but when I made the experiment with water, by filling the glass globe, in the center of which the bulb of the thermometer was suspended, with that fluid, and afterwards varied the experiment by adding 16 grains of raw silk to the water, I did not find that the conducting power of the water was sensibly impaired by the presence of the silk.*

But we have just seen that the same silk, mixed with an equal volume of air, diminished its conducting power in a very remarkable degree; consequently, there is great reason to conclude that water and air conduct Heat in a *different manner*.

But the following experiment, I think, puts the matter beyond all doubt.

It is well known that the power which air possesses of holding water in solution is augmented by Heat, and diminished by cold, and that, if hot air is saturated with water, and if this air is afterwards cooled, a part of its water is necessarily deposed.

I took a cylindrical bottle of very clear transparent glass, about 8 inches in diameter, and 12 inches high, with a short and narrow neck, and, suspending a small piece of linen rag, moderately wet, in the middle of it, I plunged it into a large vessel of water, warmed to about 100° of Fahrenheit's thermometer, where I suffered it to remain till the contained air was not only warm, but thoroughly saturated with the moisture which it attracted from the linen rag, the mouth of the bottle being well stopped up during this time with a good cork; this being done, I removed the cork for a moment, to take away the linen rag, and, stopping up the bottle again immediately, I took it out of the warm water, and plunged it into a large cylindrical jar, about 12 inches in diameter, and 16 inches high, containing just so much ice-cold water, that, when the bottle was plunged into it, and quite covered by it, the jar was quite full.

As the jar was of very fine transparent glass, as well as the bottle, and as the cold water contained in the jar was perfectly clear, I could see what passed in the bottle most distinctly; and having taken care to place the jar upon a table near the window, in a very

* The experiment here mentioned was made in the year 1787; but the result of a more careful investigation of the subject has since shown that Heat is not propagated in water in the manner here supposed. (See Essay VII, Edition of 1798.)

favourable light, I set myself to observe the appearance which should take place, with all that anxious expectation which a conviction that the result of the experiment must be decisive naturally inspired.

I was certain that the air contained in the bottle could not part with its Heat, without at the same time—that is to say, *at the same moment*, and *in the same place*—parting with a portion of its water; if, therefore, the Heat penetrated the mass of air from the center to the surface, or *passed through it* from particle to particle, in the same manner as it is probable that it passes through water, and all other unelastic fluids,* by far the greater part of the air contained in the bottle would part with its Heat, when *not actually in contact with the glass*, and a proportional part of its water being let fall at the same time, and in the *same place*, would necessarily descend in the form of rain; and, though this rain might be too fine to be visible in its descent, yet I was sure I should find it at the bottom of the bottle, if not in visible drops of water, yet in that kind of cloudy covering which cold glass acquires from a contact with hot steam or watery vapour.

But if the particles of air, instead of communicating their Heat from one to another from the center to the surface of the bottle, each in its turn, and for itself, came to the surface of the bottle, and there deposited its Heat and its water, I concluded that the cloudiness occasioned by this deposit of water would appear all over the bottle, or, at least, not more of it at the bottom than at the sides, but rather less; and this I found to be the case in fact.

The cloudiness first made its appearance upon the sides of the bottle, near the top of it; and from thence it gradually spread itself downwards, till, growing fainter as it descended lower, it was hardly visible at the distance of half an inch from the bottom of the bottle; and upon the bottom itself, which was nearly flat, there was scarcely the smallest appearance of cloudiness.

These appearances, I think, are easy to be accounted for. The

* This opinion respecting the manner in which Heat is propagated in water, and other unelastic fluids, was afterwards found to be erroneous, as has been shown in the preceding Essay.

air immediately in contact with the glass being cooled, and having deposited a part of its water upon the surface of the glass, at the same time that it communicates to it its Heat, slides downwards by the sides of the bottle in consequence of its increased specific gravity, and, taking its place at the bottom of the bottle, forces the whole mass of hot air upwards; which, in its turn, coming to the sides of the bottle, *there* deposits its Heat and its water, and afterwards bending its course downwards, this circulation is continued till all the air in the bottle has acquired the exact temperature of the water in the jar.

From hence it is clear why the first appearance of condensed vapour is near the top of the bottle, as also why the greatest collection of vapour is in that part, and that so very small a quantity of it is found nearer the bottom of the bottle.

This experiment confirmed in me an opinion which I had for some time entertained, that, though the particles of air individually, or each for itself, are capable of receiving and *transporting* Heat, yet air in a quiescent state, or as a fluid whose parts are at rest with respect to each other, is not capable of conducting it, or giving it a passage; in short, that Heat is incapable of *passing through a mass of air*, penetrating from one particle of it to another, and that it is to this circumstance that its non-conducting power is principally owing.

It is also to this circumstance, in a great measure, that it is owing that its non-conducting power, or its apparent warmth when employed as a covering for confining Heat, is so remarkably increased upon its being mixed with a small quantity of any very fine, light, solid substance, such as the raw silk, fur, eider-down, &c., in the foregoing experiments; for as I have already observed, though these substances, in the very small quantities in which they were made use of, could hardly have prevented, in any considerable degree, the air from conducting or giving a passage to the Heat, had it been capable of passing through it, yet they might very much impede it in the operation of transporting it.

But there is another circumstance which it is necessary to take into the account, and that is the attraction which subsists between air and the bodies above mentioned, and other like substances, constituting natural and artificial cloathing. For, though the incapacity of air to give a passage to Heat in the manner solid bodies permit it to pass through them may enable us to account for its warmth under certain circumstances, yet the bare admission of this principle does not seem to be sufficient to account for the very extraordinary degrees of warmth which we find in furs and in feathers, and in various other kinds of natural and artificial cloathing; nor even that which we find in snow; for if we suppose the particles of air to be at liberty to *carry off* the Heat which these bodies are meant to confine, without any other obstruction or hindrance than that arising from their *vis inertiæ*, or the force necessary to put them in motion, it seems probable that the succession of fresh particles of cold air, and the consequent loss of Heat, would be much more rapid than we find it to be in fact.

That an attraction, and a very strong one, actually subsists between the particles of air and the fine hair or furs of beasts, the feathers of birds, wool, &c., appears by the obstinacy with which these substances retain the air which adheres to them, even when immersed in water, and put under the receiver of an air-pump; and that this attraction is essential to the warmth of these bodies, I think is very easy to be demonstrated.

In furs, for instance, the attraction between the particles of air and the fine hairs in which it is concealed being greater than the increased elasticity or repulsion of those particles with regard to each other, arising from the Heat communicated to them by the animal body, the air in the fur, though heated, is not easily displaced; and this coat of confined air is the real barrier which defends the animal body from the external cold. This air cannot *carry off* the Heat of the animal, because it is itself confined, by its attraction to the hair or fur; and it transmits it with great difficulty, if it transmits at all, as has been abundantly shown by the foregoing experiments.

Hence it appears why those furs which are the finest, longest, and thickest, are likewise the warmest; and how the furs of the beaver, of the otter, and of other like quadrupeds which live much in water, and the feathers of water-fowls, are able to confine the Heat of those animals in winter, notwithstanding the extreme coldness and great conducting power of the water in which they swim. The attraction between these substances and the air which occupies their interstices is so great that this air is not dislodged even by the contact of water, but, remaining in its place, it defends the body of the animal at the same time from being wet, and from being robbed of its Heat by the surrounding cold fluid, and it is possible that the pressure of this fluid upon the covering of air confined in the interstices of the fur, or feathers, may at the same time increase its warmth, or non-conducting power, in such a manner that the animal may not, in fact, lose more heat when in water than when in air: for we have seen, by the foregoing experiments, that, under certain circumstances, the warmth of a covering is increased by bringing its component parts nearer together, or by increasing its density even at the expense of its thickness. But this point will be further investigated hereafter.

Bears, wolves, foxes, hares, and other like quadrupeds, inhabitants of cold countries, which do not often take the water, have their fur much thicker upon their backs than upon their bellies. The heated air occupying the interstices of the hairs of the animal tending naturally to rise upwards, in consequence of its increased elasticity, would escape with much greater ease from the backs of quadrupeds than from their bellies, had not Providence wisely guarded against this evil by increasing the obstructions in those parts, which entangle it and confine it to the body of the animals. And this, I think, amounts almost to a proof of the principles, assumed relative to the manner in which Heat is carried off by air and the causes of the non-conducting power of air, or its apparent warmth, when, being combined with other bodies, it acts as a covering for confining Heat.

The snows which cover the surface of the earth in winter, in high latitudes, are doubtless designed by an all-provident Creator as a garment to defend it against the piercing winds from the polar regions, which prevail during the cold season.

These winds, notwithstanding the vast tracts of continent over which they blow, retain their sharpness as long as the ground they pass over is covered with snow; and it is not till, meeting with the ocean, they acquire, from a contact with its water, the Heat which the snows prevent their acquiring from the earth, that the edge of their coldness is taken off, and they gradually die away and are lost.

The winds are always found to be much colder when the ground is covered with snow than when it is bare, and this extraordinary coldness is vulgarly supposed to be communicated to the air by the snow; but this is an erroneous opinion, for these winds are in general much colder than the snow itself.

They retain their coldness because the snow prevents them from being warmed at the expence of the earth; and this is a striking proof of the use of the snows in preserving the Heat of the earth during the winter in cold latitudes.

It is remarkable that these winds seldom blow from the poles directly towards the equator, but from the land towards the sea. Upon the eastern coast of North America the cold winds come from the northwest; but upon the western coast of Europe they blow from the northeast.

That they should blow towards those parts where they can most easily acquire the Heat they are in search of, is not extraordinary; and that they should gradually cease and die away, upon being warmed by a contact with the waters of the ocean, is likewise agreeable to the nature and causes of their motion; and if I might be allowed a conjecture respecting the principal use of the seas, or the reason why the proportion of water upon the surface of our globe is so great, compared to that of the land, it is to maintain a more equal temperature in the different climates, by heating or cooling the winds which at certain periods blow from the great continents.

That cold winds actually grow much milder upon passing over the sea, and that hot winds are refreshed by a contact with its waters, is very certain; and it is equally certain that the winds from the ocean are, in all climates, much more temperate than those which blow from the land.

In the islands of Great Britain and Ireland, there is not the least doubt but the great mildness of the climate is entirely owing to their separation from the neighbouring continent by so large a tract of sea; and in all similar situations, in every part of the globe, similar causes are found to produce similar effects.

The cold northwest winds which prevail upon the coast of North America during the winter seldom extend above 100 leagues from the shore, and they are always found to be less violent, and less piercing, as they are further from the land.

These periodical winds from the continents of Europe and North America prevail most towards the end of the month of February, and in the month of March; and I conceive that they contribute very essentially towards bringing on an early spring, and a fruitful summer, particularly when they are very violent in the month of March, and if at that time the ground is well covered with snow. The whole atmosphere of the polar regions being, as it were, transported into the ocean by these winds, is there warmed and saturated with water: and, a great accumulation of air upon the sea being the necessary consequence of the long continuance of these cold winds from the shore, upon their ceasing the warm breezes from the sea necessarily commence, and, spreading themselves upon the land far and wide, assist the returning sun in dismantling the earth of the remains of her winter garment, and in bringing forward into life all the manifold beauties of the new-born year.

This warmed air which comes in from the sea, having acquired its Heat from a contact with the ocean, is, of course, saturated with water; and hence the warm showers of April and May, so necessary to a fruitful season.

The ocean may be considered as the great reservoir and equalizer of Heat; and its benign influences in preserving a proper temperature in the atmosphere operate in all seasons and in all climates.

The parching winds from the land under the torrid zone are cooled by a contact with its waters, and, in return, the breezes from the sea, which at certain hours of the day come in to the shores in almost all hot countries, bring with them refreshment, and, as it were, new life and vigor both to the animal and vegetable creation, fainting and melting under the excessive Heats of a burning sun. What a vast tract of country, now the most fertile upon the face of the globe, would be absolutely barren and uninhabitable on account of the excessive Heat, were it not for these refreshing sea-breezes! And is it not more than probable, that the extremes of heat and of cold in the different seasons in the temperate and frigid zones would be quite intolerable, were it not for the influence of the ocean in preserving an equability of temperature?

And to these purposes the ocean is wonderfully well adapted, not only on account of the great power of water to absorb Heat, and the vast depth and extent of the different seas (which are such that one summer or one winter could hardly be supposed to have any sensible effect in heating or cooling this enormous mass); but also on account of the continual circulation which is carried on in the ocean itself by means of the currents which prevail in it. The waters under the torrid zone being carried by these currents towards the polar regions, are there cooled by a contact with the cold winds, and, having thus communicated their Heat to these inhospitable regions, return towards the equator, carrying with them refreshment for those parching climates.

The wisdom and goodness of Providence have often been called in question with regard to the distribution of land and water upon the surface of our globe, the vast extent of the ocean having been considered as a proof of the little regard that has been paid to man in this distribution. But the more light we acquire respecting the real constitution of things, and the various uses of the different parts of the visible creation, the less we shall be disposed to indulge ourselves in such frivolous criticisms.

CHAPTER 9

HEAT AS A MODE OF MOTION

ALTHOUGH Count Rumford's idea of the nature of heat was far from our modern kinetic theory, he carried out a number of experiments to demonstrate that liquids were in continuous motion even at a constant temperature. The following paper entitled "Of the Slow Progress of the Spontaneous Mixtures of Liquids" describes his method of demonstrating the existence of particle motion in a liquid at a uniform temperature by making visible the diffusion of liquids of different densities into each other.

In these experiments he was clearly trying to differentiate between convective mixtures of liquids due to temperature gradients and the as yet not clearly defined processes of diffusion at uniform temperatures. Although Rumford did not bring these experiments to any very clear conclusion, he was satisfied that he was able to demonstrate particle motion as an essential characteristic of the liquid state.

Of the Slow Progress of the Spontaneous Mixtures of Liquids Disposed to Unite Chemically with each other

Read before the National Institute of France
March 29, 1807

Mémoires de l'Institut National de France, Classe de Sciences Mathématiques et Physiques VIII, ii, pages 100–115 (1807)
Tilloch's Philosophical Magazine XXXIV, 155 (1807)

In order to obtain the most exact knowledge of the nature of the forces which act in the chemical combination of various bodies, one must study the phenomena of these operations, not only in their results, but more especially in their progress.

When we mix together two liquids which we wish to have unite, we take care to shake them violently, in order to facilitate their union; it might, however, be very interesting to know what would happen, if, instead of mixing them, they were simply brought into contact by placing one upon the other in the same vessel, taking care to cause the lighter to rest upon the heavier.

Will the mixture take place under such circumstances? and with what degree of rapidity? These are questions interesting alike to the chemist and to the natural-philosopher.

The result would depend, without doubt, on several circumstances which we might be able to anticipate, and the effects of which we might perhaps estimate *à priori*. But since the results of experiments, when they are well made, are incomparably more satisfactory than conclusions drawn from any course of reasoning, especially in the case of the mysterious operations of Nature, I propose to speak before this illustrious Assembly simply of experiments that I have performed.

Having procured a cylindrical vessel of clear white glass 1 inch 8 lines in diameter, and 8 inches high, provided with a scale divided from the bottom upwards into inches and lines, I put it on a firm table in the middle of a cellar, where the temperature, which seemed to be tolerably constant, was 64 degrees of Fahrenheit's scale.

I then poured into this vessel, with due precautions, a layer of a saturated aqueous solution of muriate of soda, 3 inches in thickness, and on to this a layer of the same thickness of distilled water. This operation was performed in such a way that the two liquids lay one upon the other without being mixed, and when everything was at rest I let a large drop of the essential oil of cloves fall into the vessel. This oil being specifically heavier than water, and lighter than the solution of muriate of soda on which the water rested, the drop descended through the layer of water; when, however, it reached the neighbourhood of the surface of the saline solution it remained there, forming a little spherical ball, which maintained its position at rest, as though it were suspended, near the axis of the vessel.

I then poured, with proper precautions, a layer of olive oil four lines in thickness on to the surface of the water, to prevent the contact of the air with the liquid, and having observed, by means of the scale attached to the vessel, and noted down in a register, the height at which the little ball was suspended, I withdrew, and, locking the door, I left the apparatus to itself for twenty-four hours.

In a preliminary experiment, made to determine in what proportions the saturated solution should be mixed with distilled water, that the mixture might have the same specific gravity as the oil of cloves, I found that a mixture composed of 1 measure of the solution and 9 measures of distilled water had a slightly higher specific gravity than the oil; but with 10 measures of distilled water the oil sank in the mixture.

As the little ball of oil, designed to serve me as an index, was suspended a very little above the upper surface of the layer of the saturated solution, this showed me that the precautions which I had taken were sufficient to prevent the mixing of the distilled water and the saline solution when I put one upon the other, and I knew that this mixture could not take place subsequently without causing at the same time my little *sentinel*, which was there to warn me of this event, to ascend.

There was, however, a single source of error which I was obliged to guard against. I had observed, in other experiments of this kind, that the air which was disseminated through or dissolved in water containing in solution a small quantity of muriate of soda left the liquid, and attached itself to the little ball of oil of cloves which I had introduced into it, and, having formed on top of it a little bubble scarcely visible, caused it to ascend in the liquid, even when the density of the liquid had not changed at all.

To prevent this accident, I boiled for some time both the saturated solution and the distilled water employed in the experiment, in order to free them from air, and, for the same reason, I subsequently covered the water with a layer of olive oil to prevent the contact of this water with the atmospheric air.

After the little apparatus mentioned above had been left to itself for twenty-four hours, I entered the cellar, taking a light in order to note the progress of the experiment, and I found that the little ball had risen 3 lines.

The next day, at the same hour, I observed the ball again, and I found that it had risen about 3 lines more; and this it continued to ascend about 3 lines a day for six days, when I put an end to the experiment.

I afterwards made nearly similar experiments with saturated aqueous solutions of nitrate of potash, carbonate of potash, and carbonate of soda. In each of these experiments the surface of the saturated solution was covered with a layer of distilled water 3 inches in thickness, but the surface of this layer of water was not covered by a layer of olive oil; it was exposed to the air, and this circumstance was, without doubt, the reason that the daily results of a single experiment were not always the same two days in succession.

The little ball of oil of cloves, which served as an index to mark the progress of the mixture of the saturated solution with the distilled water resting upon it, ascended usually 2 or 3 lines in twenty-four hours, but sometimes I found that it had left its position and had risen to the very surface of the water.

In such cases it was, without doubt, borne upwards by the air which it had attracted from the liquid; for when I allowed a fresh drop of the same oil to fall into the water, I found that it never failed to descend immediately in the liquid, and to take up its position 2 or 3 lines above the level at which, the day before, I had found the ball which had now left its place.

In the experiments made with solutions of carbonate of soda and carbonate of potash, the balls of oil changed in appearance by the end of two or three days; from being transparent, they became semi-opaque and of a whitish color; they changed at the same time with regard to their specific gravity as well, and became a little lighter. These changes were evidently due to the beginning of saponification.

This accidental circumstance made it necessary for me to renew each day the drop of oil which served as the index, allowing the others to pursue their way to the surface of the liquid without paying any further attention to them.

By using as indices little glass balloons of proper size and thickness, instead of the drops of oil, the inconveniences arising from the saponification of the oil might be avoided.

But without spending more time on the details of these experiments, I hasten to return to their results. They showed that the mixture went on continually, but very slowly, between the various aqueous solutions employed and the distilled water resting upon them.

There is nothing in this result to excite the surprise of any one, especially of chemists, unless it is the extreme slowness of the progress of the mixture in question. The fact, however, gives occasion for an inquiry of the greatest importance, which is far from being easy to solve.

Does this mixture depend upon a peculiar force of attraction different from the attraction of universal gravitation, a force which has been designated by the name of chemical affinity? Or is it simply a result of motions in the liquids in contact, caused by changes in their temperatures? Or is it, perhaps, the result of a peculiar and continual motion common to all liquids, caused by the instability of the equilibrium existing among their molecules?

I am very far from assuming to be able to solve this great problem, but it has often been the subject of my thoughts, and I have made at different times a considerable number of experiments with a view of throwing light into the profound darkness with which the subject is shrouded on every side.

CHAPTER 10

RADIATION

THE transmission of heat by means of radiation was a real puzzle to eighteenth-century physicists. Many believed that heat actually existed in two quite different forms and many of the caloricists excluded radiant heat from their fluid theories. It was also a common point of view, which was shared by Count Rumford, that heat or cold were two distinctly different physical phenomena.

As a result of the lack of agreement on the nature of radiant heat, a great many natural philosophers turned their attention toward designing instruments and experiments to separate the various apparent forms of heat and to shed some light on the physics involved. The following paper exemplifies Rumford's method of approach and clearly shows his understanding of many of the laws of radiation.

Since so many physicists around the turn of the nineteenth century were trying to invent apparatus for studying radiant heat, it is not surprising that simultaneous inventions resulted. The interesting controversy between Leslie and Rumford about radiant-heat experiments serves as an illustration of this kind of duplication and shows the lively interest the educated public had in scientific disputes.

An Inquiry concerning the Nature of Heat, and the Mode of its Communication

Read before the Royal Society, February 2, 1804

Philosophical Transactions XCIV, 77–182 (1804)
Bibliotheque Britannique (Sciences et Arts) XXV, 185–221, 273–311 (1804)
Nicholson's Journal IX, 58–63, 193–203 (1804)

Heat is employed in such a vast variety of different processes, in the affairs of life, that every new discovery relative to it must necessarily be of real importance to mankind; for, by obtaining a more intimate knowledge of its nature and mode of action, we shall no doubt be enabled not only to excite it with greater economy, but also to confine it with greater facility, and direct its operations with more precision and effect.

Having many years ago found reason to conclude that a careful observation of the phenomena which attend the heating and cooling of bodies, or the communication of heat from one body to another, would afford the best chance of acquiring a farther insight into the nature of heat, my view, in all my researches on this subject, has been principally directed to that point; and the experiments of which I am now to give an account may be considered as a continuation of those I have already, at different times, had the honour of laying before the Royal Society, and of presenting to the public in my Essays.

In order that the attention of the Society may not be interrupted unnecessarily by description of instruments in the midst of the accounts of interesting experiments, I shall begin by describing the apparatus which was provided for these researches; and, as a perfect knowledge of the instruments made use of is indispensably necessary in order to form distinct ideas of the experiments, I shall take the liberty to be very particular in these descriptions.

This instrument, which I shall take the liberty to call a *thermoscope*, is very simple in its construction. Like the hygrometer of Mr. Leslie (as he has chosen to call his instrument) it is composed of two glass balls, attached to the two ends of a bent glass tube; but the balls, instead of being near together, are placed at a considerable distance from each other; and the tube which connects them, instead of being bent in its middle, and its two extremities turned upwards, is quite straight in the middle, and its two extremities, to which its two balls are attached, are turned perpendicularly upwards, so as to form each a right angle with the middle part of the tube, which remains in a horizontal position.

At one of the elbows of this tube there is inserted a short tube of nearly the same diameter, by means of which a very small quantity of spirit of wine, tinged of a red colour, is introduced into the instrument; and, after this is done, the end of this short tube (which is only about an inch long) is sealed hermetically; and all communication is cut off between the air in the balls of the instrument and in its tube and the external air of the atmosphere.

A small *bubble* of the spirit of wine (if I may be allowed to use that expression) is now made to pass out of the short tube into the long connecting tube; and the operation is so managed that this bubble (which is about $\frac{3}{4}$ of an inch in length) remains stationary, at or near the middle of the horizontal part of the tube, *when the temperature (and consequently the elasticity) of the air in the two balls, at the two extremities of the tube, is precisely the same.*

By means of a scale of equal parts, attached to the horizontal part of the connecting tube, the position of the bubble can be ascertained, and its movements observed.

If now, the bubble being at rest in its proper place, one of the balls of the instrument be exposed to the calorific rays which proceed in all directions from a hot body, while the other ball is defended from those rays by a screen, the air in the ball so exposed to the action of these rays will be heated; and, its elasticity being increased by this additional heat, its pressure will no longer be counterbalanced by the elasticity of the colder air in the other ball, and the bubble will be forced to move out of its place and to take its station nearer to the colder ball.

By presenting two hot bodies at the same time to the two balls of the instrument, taking care that each ball shall be defended from the action of the hot body presented to the opposite ball, the distances of these hot bodies from their respective balls may be so regulated that their actions on those balls may be equal, however the temperature of those hot bodies may differ, or however different may be the quantities or intensities of the calorific rays which they emit.

The instrument will show, with the greatest certainty, when the actions of these hot bodies on their respective balls are equal; for, until they become *unequal*, the bubble will remain immovable in its place.

And, when the actions of two hot bodies on the instrument are equal, the relative intensities of the rays they emit may be ascertained by the distances of the bodies from the balls of the instrument.

If their distances from their respective balls are equal, the intensities of the rays they emit must, of course, be equal.

If those distances are unequal, the intensities will probably be as the squares of the distances, inversely.

A distinct and satisfactory idea may be formed of the instrument I have been describing, from Fig. 2 [Plate 10].

AB is a board, 27 inches long, 9 inches wide, and 1 inch thick, which serves as a support for the bent tube CDE, at the two extremities of which the two balls are fixed. The two projecting ends of the tube, C and E, which are in a vertical position, are each 10 inches long; and the horizontal part D of the tube, which is fastened down on the board, is 17 inches in length.

The balls are each 1.625 inches in diameter. The diameter of the tube is such, that 1 inch of it in length would contain 15 grains Troy of mercury.

The pillar F, which, by means of a horizontal arm projecting from it, serves for supporting the circular vertical screen represented in the figure, is firmly fixed in the board AB.

This circular screen (which is made of pasteboard, covered on both sides with gilt paper) serves for preventing one of the balls of the instrument from being affected by the calorific rays proceeding from a hot body which is presented to the opposite ball.

Besides the circular screen represented in the figure, several other screens are used in making experiments; for the instrument is so extremely sensible, that the naked hand presented to one of the balls, at the distance of several inches, puts the bubble in motion; and it is affected very sensibly by the rays which proceed from the person who approaches it to make the experiments, unless care be taken, by the interposition of screens, to prevent those rays from falling on the balls. These screens can be best and most readily made by providing light wooden frames, about two feet square, and half an inch in thickness, and covering them on both sides, first with thick cartridge paper, and then with what is called

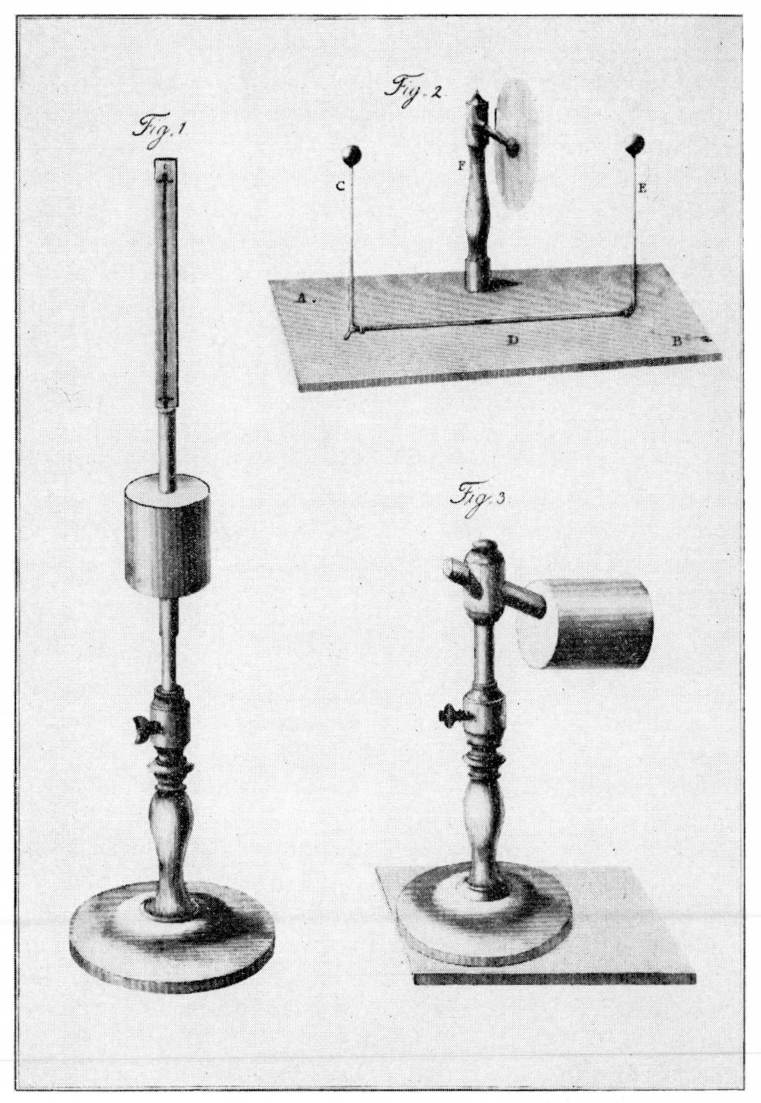

PLATE 10. Count Rumford's equipment for measuring radiation phenomena.

gilt paper; the metallic substance (copper) with which one side of the paper is covered being on the outside.

To support a movable screen of this kind in a vertical position, it must of course be provided with a foot or stand. Those I use are fastened to one side of a pillar of wood by two screws, one of which passes through the centre of the screen where the cross-bars belonging to the frame of the screen meet, and the other through the middle of the piece of wood which forms the bottom of the screen. This pillar of wood, which is turned in a lathe, is 12½ inches high, and is firmly fixed, at its lower end, in a piece of wood 8 inches square and 1 inch thick, which serves as a stand or foot for supporting it.

As, in making experiments with this *thermoscope*, it is frequently necessary to remove the hot bodies that are presented to it farther from it or to bring them nearer to it, in order that this may be done easily and expeditiously by one person, and without its being necessary for him to remove his eye from the bubble (which he should constantly have in his view), I make use of a simple machine, which I have found to be very useful.

It is a long and shallow wooden box, open at both ends. It is 6 feet long, 12 inches wide, and 5 inches deep, measured on the outside; its vertical sides are made of 1½ inch deal; its bottom and top, of inch deal. A part only of the top or cover of this box is fixed down on the sides, and is immovable. The part of the cover which is fixed, and on which the thermoscope is placed, occupies the middle of the box, and is 13 inches in length. On the right and left of this fixed part, the top of the box is covered by a sliding board, 2 feet 3 inches long, which passes in deep grooves, made to receive it, in the sides of the box. A rack is fixed to the under side of each of these sliding boards; and there is a small cog wheel in the box, the axis of which passes through the sides of the box, and is furnished with a winch in the front of the box. By turning round these wheels by means of their winches (both of which can be managed by the same person, at the same time), the sliders may be moved backwards and forwards at pleasure.

In order to ascertain with facility and dispatch the distances of the hot bodies from their respective balls, the top of the front side of the wooden box is divided into inches on each side of the fixed part of the cover of the box; and there is a *nonius* belonging to each of the sliders, which is placed in such a manner as to indicate, at all times, the exact distance of the hot body from its corresponding ball.

The level of the upper surface of that part of the cover which is fixed is about $\frac{1}{8}$ of an inch higher than the level of the upper surface of the sliders, in order that, when a thermoscope longer than this fixed part is placed on it, the sliders may pass freely under its two projecting ends without deranging it.

It is evident, from this description, that by placing the thermoscope on the fixed part of the cover of the box, with its two balls in a line parallel to the axis of the box, and by placing the two hot bodies presented to the two balls of the instrument (elevated to a proper height) on stands set down on the sliders, an observer, by taking the two winches in his hands, keeping his eye fixed on the bubble, may, with the greatest facility, so regulate the distances of the hot bodies from their respective balls that the bubble shall remain immovable in its place.

In order to be able to ascertain precisely the temperatures of the hot bodies presented to this instrument, and in order that their surfaces might be equal, two equal cylindrical vessels, of thin sheet brass, with oblique cylindrical necks, were provided, of the form represented in Figure 3 [Plate 10].

This cylindrical vessel, which is placed in a horizontal position in order that its flat bottom may be presented *in a vertical position* to one of the balls of the thermoscope, is so fixed to a wooden stand, of a peculiar construction, that it may be raised or lowered at pleasure. This is necessary, in order that its axis may be in the continuation of a line passing through the centres of the two balls of the thermoscope.

This cylindrical vessel is 3 inches in diameter and 4 inches in length, and its oblique cylindrical neck is 0.86 of an inch in diameter and 3.8 inches in length.

The neck of this vessel is inserted *obliquely* into its cylindrical body, in order that the water with which it is occasionally filled may not run out of it, when the body of the vessel is laid down in a horizontal position, in the manner represented in the above-mentioned figure.

A thermometer, with a cylindrical bulb 4 inches in length, being inserted into the body of this vessel, through its neck, shows the temperature of the contained water.

Care is necessary, in constructing a thermoscope, to choose a tube of a proper diameter; if its bore be too small, it will be found very difficult to keep the spirit of wine in one mass; and if it be too large, the little horizontal column it forms (which I have called a bubble) will be ill defined at its two ends, which will render it difficult to ascertain its precise situation. After a number of trials I have found that a tube, the bore of which is of such a size that 1 inch of it in length contains about 15 or 18 grains Troy of mercury, answers best. For a tube of that size the balls may be about $1\frac{1}{2}$ inch in diameter; and they should both be painted black with Indian ink, which renders the instrument more sensible.

I have an instrument of this kind, the tube of which is quite filled with spirit of wine, excepting only the space occupied by a small bubble of air, which is introduced into the middle of the horizontal part of the tube; but it does not answer so well as those which contain only a very small quantity of that liquid, sufficient to form a small bubble.

But, without enlarging any farther, at present, on the construction of these instruments, I now proceed to give an account of the experiments for which they were contrived.

Having found abundant reason to conclude, from the results of the experiments of which an account has already been given, that all the heat which a hot body loses when it is exposed in the air to cool is not given off to the air which comes into contact with it, but that a large proportion of it escapes in rays, which do not heat the transparent air through which they pass, but, like light, generate heat only when and where they are stopped and absorbed, —I suspected that in every case when, in the foregoing experiments, the cooling of my instruments was expedited by coverings applied to their metallic surfaces, those coverings must, by some means or other, have facilitated and accelerated the emission of calorific rays from the hot surface.

Those suspicions implied, it is true, the supposition that different substances, heated to the same temperature, emit unequal quantities of calorific rays; but I saw no reason why this might not be the case in fact; and I hastened to make the following experiments, which put the matter beyond all doubt.

Experiment No. 12.—Two equal cylindrical vessels, made of sheet brass, and polished very bright, each 3 inches in diameter and 4 inches long, suspended by their oblique necks in a horizontal position (being placed on their wooden stands), were filled with water at the temperature of 180°; and their circular flat bottoms were presented in a vertical position to the two balls of the thermoscope, at the distance of 2 inches.

When the two hot bodies were presented, at the same moment, to the two balls of the instrument, or, what was still better, when two screens were placed before the two balls, at the distance of about an inch, and, after the hot bodies were placed, these screens were both removed at the same instant, the small column of spirit of wine, which I have called a *bubble*, remained immovable in its place, in the middle of the horizontal part of the tube of the instrument.

If one of the hot bodies was now brought nearer the ball to which it was presented (the other hot body remaining in its place), the bubble immediately began to move from the hot body which was advanced forward, towards the opposite ball to which the other hot body was presented.

If, instead of advancing one of the hot bodies nearer the ball to which it was presented, it was drawn backward to a greater distance from it, the action of its calorific rays on the ball was diminished by this increase of distance; and, being overcome by the action of the rays from the hot body presented to the opposite ball (at a smaller distance), the bubble was forced out of its place, and obliged to move towards the ball which had been drawn backward.

When one of the hot bodies only was presented to one of the

balls, the bubble was immediately put in motion, and by bringing the hot body nearer to the ball, it might be driven quite out of the tube into the opposite ball; this, however, should never be done, because it totally deranges the instrument, as it is easy to perceive it must do.

Having, by these trials, ascertained the sensibility and the accuracy of my instrument, I now proceeded to make the following decisive experiment.

Experiment No. 13.—Having blackened the flat circular bottom of one of the cylindrical vessels by holding it over the flame of a wax candle, I now filled both vessels again with water at the temperature of 180°F., and presented them, as before, to the two opposite balls of the instrument at equal distances.

The bubble was instantly driven out of its place by the superior action of the blackened surface, and did not return to its former station till after the vessel which was blackened had been removed to more than 8 inches from the ball to which it was presented; the other vessel, which had not been blackened, remaining in its former situation, at the distance of 2 inches from its ball.

The result of this experiment appeared to me to throw a new light on the subject which had so long engaged my attention, and to present a wide and very interesting field for farther investigation.

I could now account, in a manner somewhat more satisfactory, for those appearances in the foregoing experiments which were so difficult to explain,—for the acceleration of the passage of the heat out of my instruments, which resulted from covering them with linen, varnish, &c.; and I immediately set about making a variety of new experiments, from which I conceived I should acquire a farther insight into those invisible mechanical operations which take place when bodies are heated and cooled.

Finding so great a difference in the quantities of calorific rays which are thrown off by the polished surface of a metal when exposed *naked* to the cold air and when *blackened*, I now proceeded to make experiments to ascertain whether or not all those substances with which the sides of my cylindrical vessels had been covered, and which had been found to expedite the cooling of those instruments, would also facilitate the emission of calorific rays from the surfaces of the instruments I presented to the balls of my thermoscope; and I found this to be the case in fact.

As the results of all these experiments proved, in the most decisive manner, that all the substances which, when applied to the metallic surfaces of my large cylindrical vessels, had expedited their cooling, facilitated and expedited the mission of calorific rays, I could no longer entertain any doubts respecting the agency of *radiation* in the heating and cooling of bodies. Many important points, however, still remained to be investigated before distinct and satisfactory ideas could be formed respecting the nature of those rays and the mode of their action.

I had hitherto made use of but one metal (brass) in my experiments; and that was not a simple, but a compound metal. The first subject of inquiry which presented itself, in the prosecution of these researches, was to find out whether or not similar experiments made with other metals would give similar results.

Experiment No. 14.—Procuring from a gold-beater a quantity of leaf gold and leaf silver about three times as thick as that which is commonly used by gilders, I covered the surfaces of the two large cylindrical vessels, No. 1 and No. 2, with a single coating of oil varnish; and, when it was sufficiently dry for my purpose, I gilt the instrument No. 1 with the gold leaf, and covered the other, No. 2, with silver leaf. When the varnish was perfectly dry and hard, I wiped the instruments with cotton, to remove the superfluous particles of the gold and silver, and then repeated the experiment, so often mentioned, of filling the instruments with boiling-hot water, and exposing them to cool in the air of a large quiet room.

The time of cooling through the given interval of 10 degrees was just the same as it was before, when the natural surface of these brass vessels was exposed *naked* to the air. I repeated the experiment several times, but could not find that the difference in the metals made any difference in the times of cooling.

Experiment No. 15.—Not satisfied to rest the determination of so important a point on a trial with three metals only,—brass, gold, and silver,—I now provided myself with two new instruments,—the one made of lead, and the other covered with tinned sheet-iron, improperly, in England, called tin.

As the *conducting power* of lead, with respect to heat, is much greater than that of any other metal, I conceived that, if the *radiation* of a body were any way connected with its *conducting power*, the cooling of the water contained in the leaden vessel would necessarily be either more or less rapid than in a vessel constructed of any other metal.

The result of this experiment, as also the results of several others similar to it, showed that heat is given off with the same facility, or with the same celerity, from the surfaces of all the metals.

Is not this owing to their being all equally wanting in *transparency*? And does not this afford us a strong presumption that heat is in all cases excited and communicated by means of radiations, or *undulations*, as I should rather choose to call them?

I am sensible, however, that there is another and most important question to be decided before these points can be determined; and that is, whether bodies are cooled in consequence of the rays they emit or by those they receive.

The celebrated experiment of Professor Pictet, which has often been repeated, appears to me to have put the fact beyond all doubt, that rays, or emanations, which, like light, may be concentrated by concave mirrors, proceed from cold bodies; and that these rays, when so concentrated, are capable of affecting, in a manner perfectly sensible, a delicate air thermometer.

One of the objects I had principally in view, in contriving the before-described instrument, which I have called a thermoscope, was to investigate the nature and properties of those emanations, and to find out, if possible, whether they are not of the same nature as those calorific rays which have long been known to proceed from hot bodies.

My first attempts, in these investigations, were to ascertain the existence of those emanations universally, and to discover what visible effects they might be made to produce independently of concentration by means of concave mirrors.

Experiment No. 16.—My two horizontal cylindrical vessels of sheet brass (of the same form and dimensions), having been made very clean and bright, were fixed to their stands; and, being elevated to a proper height to be presented to the balls of the thermoscope, were set down near that instrument (which was placed on a table in a large quiet room), where they were suffered to remain several hours, in order that the whole of this apparatus might acquire precisely the same temperature.

Daylight was excluded by closing the window-shutters; and, in order that the thermoscope might not be deranged by the calorific rays proceeding from the person of the observer on his entering the room to complete the intended experiments, screens were previously placed before the instrument in such a manner that its balls were completely defended from those rays.

Things having been this prepared, I entered the room as gently as possible, in order not to put the air of the room in motion, and, approaching the thermoscope, presented first one and then the other cylindrical vessel to one of the balls of the instrument; but it was not in the least degree affected by them, the bubble of spirit of wine remaining immovably in the same place.

Experiment No. 17.—Having assured myself, by these previous trials, that the instrument was not sensibly affected by a bright metallic surface being presented to it, provided the temperature of the metal and that of the instrument were the same, I now withdrew one of the cylindrical vessels, and, taking it into another room, I filled it with pounded ice and water.

Entering the room again, I now presented the flat vertical bottom of this horizontal cylindrical vessel, filled with ice and water, to one of the balls of the thermoscope at the distance of four inches.

The bubble of spirit of wine began instantly to move with a slow, regular motion towards the cold body; and, having advanced in the tube about an inch, it remained stationary.

On bringing the cold body nearer the ball to which it was presented, the bubble was again put in motion, and advanced still farther towards the cold body.

Experiment No. 18.—Although the result of the foregoing experiment appeared to me to afford the most indisputable proof of the *radiation* of cold bodies, and that the rays which proceed from them have a power of *generating cold* in warmer bodies which are exposed to their influence, yet in a matter so extremely curious, and of such high importance to the science of heat, I was not willing to rest my inquiries on the result of a single experiment.

In order to vary the substance, or species of matter, presented cold to the instrument, and at the same time to remove all suspicion respecting the possibility of the effects observed being produced by currents of cold air occasioned in the room by the presence of the cold body, I now repeated the experiment with the following variations.

The thermoscope was laid down on one side, so that the two ends of its tube, to which its balls were attached, instead of being vertical, were now in a horizontal position; and the cold body, instead of being presented to the ball of the instrument on one side of it, and on the same horizontal level with it, was now placed *directly under it*, and at the distance of 6 inches.

This cold body, instead of being a metallic substance, was a solid cake of ice, circular, flat, and about 3 inches thick, and 8 inches in diameter. It was placed in a shallow earthen dish, about 9 inches in diameter below, 12 inches in diameter above, at its brim, and 4 inches deep. The cake of ice being laid down on the bottom of the dish, the top of the dish was covered by a circular piece of thick paper, 14 inches in diameter, which had a circular hole in its centre, just 6 inches in diameter.

This earthen dish, containing the ice, and thus covered, was placed perpendicularly under one of the balls of the thermoscope, at such a distance that the centre of the upper surface of the flat cake of ice was 6 inches below the ball.

The result of this experiment was just what might have been expected: the ice was no sooner placed under the ball of the instrument than the bubble of spirit of wine began to move towards that side where the cold body was placed; and it did not remain stationary till after it had advanced more than an inch in the tube.

Experiment No. 19.—Desirous of discovering whether the surface of a liquid emits frigorific or calorific rays, as solid bodies have been found to do, I now removed the cake of ice from the earthen dish, and replaced it with an equal mass of ice-cold water.

The result of this experiment was, to all appearance, just the same as that of the last. The bubble moved towards the cold body, and took its station in the same place where it had remained stationary before. I found reason, however, to conclude, after meditating on the subject, that although the last experiment proves, in a most decisive manner, that radiations actually proceed from the surface of *water*, yet the proof of the radiation from the surface of ice, afforded by the preceding experiment, is not equally conclusive; for, as the temperature of the air of the room in which these experiments were made was many degrees above the freezing point, it is possible, and even probable, that the surface of the ice was actually covered with a very thin, and consequently invisible, coating of water during the whole of the time the experiment lasted.

Finding reason to conclude that frigorific rays are always emitted by cold bodies, and that these emanations are very analogous to the calorific rays which hot bodies emit, I was impatient to discover whether all cold bodies, at the same temperature, emit the same quantity of rays, or whether (as I had found to be the case with respect to the calorific rays emitted by hot bodies) some substances emit more of them and some less.

With a view to the ascertaining of this important point, I made the following experiments.

Experiment No. 20.—Having found that a metallic surface, rendered quite black by holding it over the flame of a wax candle, emits a much larger quantity of calorific rays when hot, than the same metal, at the same temperature, throws off when naked, I was very curious to find out whether blackening the surface of a cold metal would or would not increase, in like manner, the quantity of frigorific rays emitted by it.

Having blackened, in the manner already described, the flat bottom, or rather end, of one of my horizontal cylindrical brass vessels with an oblique neck, I filled it with a mixture of ice and common salt; and, filling another vessel of the same kind, the bottom of which was not blackened, with the same cold mixture, I presented them both, at the same instant, and at the same distance, to the two opposite balls of my thermoscope.

The result of this experiment was perfectly conclusive: the bubble of spirit of wine began immediately to move towards the ball to which the *blackened* cold body was presented; indicating thereby that that ball was more cooled by the frigorific rays which proceeded from the blackened surface than the opposite ball was cooled by the rays which proceeded from an equal surface of naked metal, at the same temperature.

As this experiment appeared to me to be of great importance, I repeated it several times, and always with the same results; the motion of the bubble, which constituted the index of the instrument, constantly showing that the frigorific rays from the blackened surface were more powerful in generating cold than those which proceeded from the naked metal.

Reserving for a future communication an account of the sequel of my inquiries respecting the subject which I have undertaken to investigate, I shall conclude this long paper with some observations concerning the *practical uses* that may be derived from a knowledge of the facts which have been established by the results of the foregoing experiments.

In all cases where it is designed to *preserve the heat* of any substance which is confined in a metallic vessel, it will greatly contribute to that end if the external surface of the vessel be very clean and bright; but if the object be to *cool* anything quickly in a metallic vessel, the external surface of the vessel should be painted, or covered with some of those substances which have been found to emit calorific rays in great abundance.

Polished tea-urns may be kept boiling hot with a much less expense of spirit of wine (burnt in a lamp under them) than such as are varnished; and the cleaner and brighter the dishes and covers for dishes are made, which are used for bringing victuals on the table, and for keeping it hot, the more effectually will they answer that purpose.

Saucepans and other kitchen utensils which are very clean and bright on the outside may be kept hot with a smaller fire than such as are black and dirty; but the bottom of a saucepan or boiler should be blackened, in order that its contents may be made to boil quickly, and with a small expense of fuel.

When kitchen utensils are used over a fire of sea-coal or of wood, there will be no necessity for blackening their bottoms, for they will soon be made black by the smoke; but when they are used over a clear fire made with charcoal, it will be advisable to blacken them,—which may be done in a few moments by holding them over a wood or coal fire, or over the flame of a lamp or candle.

Proposals have often been made for constructing the broad and shallow vessels (flats), in which brewers cool their wort, of metal, on a supposition that the process of cooling would go on faster in a metallic vessel than in a wooden vessel; but this would not be found to be the case in fact, a metallic surface being ill calculated for expediting the emission of calorific rays.

The great thickness of the timber of which brewers' flats are commonly made is a circumstance very favourable to a speedy cooling of the wort; for, when the flats are empty, this mass of wet wood is much cooled, not only by the cold air which passes over it, but also and more especially by evaporation; and when the flat is again filled with hot wort a great part of the heat of that liquid is absorbed by the cold wood.

In all cases where metallic tubes filled with steam are used for warming rooms or for heating drying-rooms, the external surface of those tubes should be painted or covered with some substance which facilitates the emission of calorific rays. A covering of thin paper will answer that purpose very well, especially if it be black, and if it be closely and firmly attached to the surface of the metal with glue.

Tubes which are designed for *conveying* hot steam from one place to another should either be well covered up with *warm* covering or should be kept clean and bright. It would, I am persuaded, be worth while, in many cases, to gild them, or at least to cover them with what is called gilt paper, or with tin foil, or some other metallic substance which does not easily tarnish in the air.

The cylinders and principal steam-tubes of steam-engines might be covered first with some warm clothing, and then with thin sheet brass kept clean and bright. The expense of this covering would, I am confident, be amply repaid by the saving of heat and fuel which would result from it.

If garden walls painted black acquire heat faster when exposed to the sun's direct rays than when they are not so painted, they will likewise cool faster during the night; and gardeners must be best able to determine whether these rapid changes of temperature are, or are not, favourable to fruit-trees.

Black clothes are well known to be very warm in the sun; but they are far from being so in the shade, and especially in cold weather. No coloured clothing is so cold as black when the temperature of the air is below that of the surface of the skin, and when the body is not exposed to the action of calorific rays from other substances.

It has been shown that the warmth of clothing depends much on the *polish* of the surface of the substance of which it is made; and hence we may conclude that, in choosing the colour of our winter garments, those dyes should be avoided which tend most to destroy that polish; and, as a white surface reflects more light than an equal surface, equally polished, of any other colour, there is much reason to think that white garments are warmer than any other in cold weather. They are universally considered as the coolest that can be worn in very hot weather, and especially when a person is exposed to the direct rays of the sun; and if they are well calculated to reflect calorific rays in summer, they must be equally well calculated to reflect those frigorific rays by which we are cooled and annoyed in winter.

I have found, by direct and decisive experiments (of which an account will hereafter be given to this Society), that garments of fur are much warmer in cold weather when worn with the fur or hair outwards than when it is turned inwards. Is not this a proof that we are kept warm by our clothing, not so much by confining our heat as by keeping off those frigorific rays which tend to cool us?

The fine fur of beasts, being a highly polished substance, is well calculated to reflect those rays which fall on it; and if the body were kept warm by the rays which proceed from it being reflected back upon it, there is reason to think that a fur garment would be warmest when worn with the hair inwards; but if it be by reflecting and turning away the frigorific rays from external (colder) bodies that we are kept warm by our clothes in cold weather, we might naturally expect that a pelisse would be warmest when worn with the hair outwards, as I have found it to be in fact.

The point here in question is by no means a matter of small importance; for until the principles of the warmth of clothing be understood, we shall not be able to take our measures with certainty, and with the least possible trouble and expense, for defending ourselves against the inclemencies of the seasons, and making ourselves comfortable in all climates.

The fur of several delicate animals becomes white in winter in cold countries, and that of the bears which inhabit the polar region is white in all seasons. These last are exposed alternately, in the open air, to the most intense cold and to the continual action of the sun's direct rays during several months. If it should be true that heat and cold are excited in the manner above described, and that white is the colour most favourable to the reflection of calorific and frigorific rays, it must be acknowledged, even by the most determined sceptic, that these animals have been exceedingly fortunate in obtaining clothing so well adapted to their local circumstances.

The excessive cold which is known to reign, in all seasons, on the tops of very high mountains and in the higher regions of the atmosphere, and the frosts at night which so frequently take place on the surface of the plains below in very clear and still weather in spring and autumn, seem to indicate that frigorific rays arrive continually at the surface of the earth from every part of the heavens.

May it not be by the action of these rays that our planet is cooled continually, and enabled to preserve the same mean temperature for ages, notwithstanding the immense quantities of heat that are generated at its surface, by the continual action of the solar rays?

If this conjecture should be well founded, we should be led to conclude that the inhabitants of certain hot countries who sleep at night on the tops of their houses, in order to be more cool and comfortable, do wisely in choosing that situation to pass their hours of rest.

DISCOVERY OF THE DIFFERENTIAL THERMOMETER*

When two scientists working in different places on similar problems manage almost simultaneously to invent similar new devices of obvious significance, a storm is sure to follow. The history of invention and of inventors is enlivened by several spectacular priority controversies which have never been settled because both claimants seem to deserve full credit.

Since the discovery is more important to the advance of knowledge than the name of the discoverer, a priority struggle of 150 years ago is interesting mostly as a comedy of manners, however inconvenient a biographer may find it. No physicist really cares today whether the differential thermometer was invented by Sir John Leslie or Count Rumford, or whether Sir Humphrey Davy did justice to Leslie in a subsequent argument, but the story of the dispute provides amusing glimpses of famous personalities of the time. One concludes that, however great the progress of science, human nature has not been changing very much.

The debate over who first produced the differential thermometer continues today, although the heat of battle has long since died away. On the European continent, Rumford is given the credit with no mention of Leslie, but English and American historians of science consider Leslie the inventor, with no reference to Rumford.

As to the facts in the case, Leslie in Scotland and Rumford in Bavaria not only invented similar thermometers at the same time, but designed them to carry out almost identical experiments on the radiation from reflecting and absorbing surfaces. So nearly identical were their investigations that immediate accusations of plagiarism were leveled against both men from various quarters. (Actually Leslie's "differential thermometer" and Rumford's "thermoscope" were not precisely identical, as Plates 11 and 10 will show. Leslie's device had a long liquid column separating the air bulbs, while Rumford's had only a small bubble of liquid running in a horizontal section of tube.)

Count Rumford gives us a very detailed account of his side of the story:†

"I performed these experiments in Munich, in 1803, during the months of

* Reproduced, with permission, from the *American Journal of Physics* **22**, 13–17 (1954).

† *The Works of Rumford* (American Academy of Arts and Sciences, Boston, 1873), vol. II, p. 188.

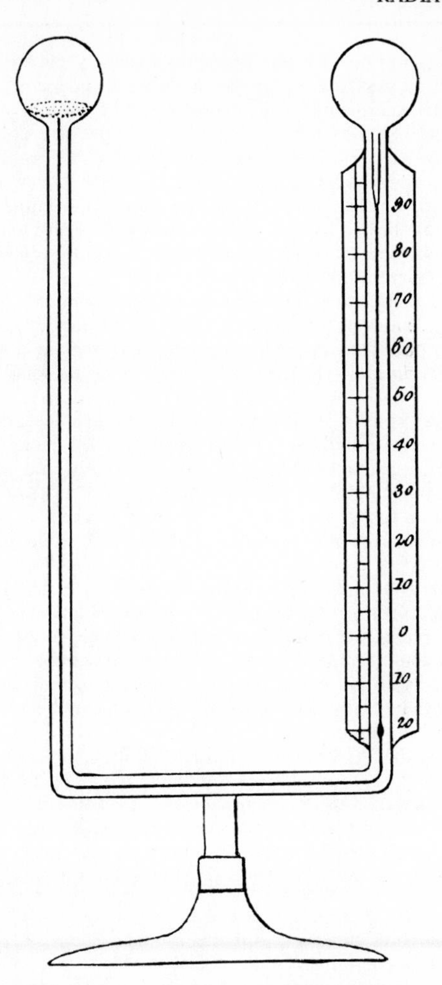

PLATE 11. Leslie's differential thermometer.

January, February, and March. According as the results seemed of importance, I immediately acquainted my friends in England and France with them. Among others, I communicated to Sir Joseph Banks, then President of the Royal Society of London, the very striking results of an experiment, on the 11th of March . . . with various vessels *blackened* and *covered with repeated coatings of varnish*, and I announced the results obtained. I also informed him of the discovery . . . of my *thermoscope.*

"Since Sir Joseph showed my letters to various persons, and since I did not keep my experiments or their results secret from him or from anyone else, my discovery was publicly mentioned in London even as early as the spring of the past year [1803]. As an incontrovertible proof of this fact, I can bring forward a letter from a friend of mine . . . in which he congratulates me on the success of my researches, and informed me at the same time that he had learned what he knew with regard to my discoveries from Mr. Davy, a Professor at the Royal Institution, who had spoken publicly of them in his lectures on chemistry.

"The 6th of June there were sent to me from London . . . a letter from Mr. Davy . . . in which he informed me that Mr. Leslie had, a short time previously, published a memoir on heat, and that in it had described various experiments which bore a resemblance to some which I had performed.

"As I had, only a short time before, occupied the attention of the National Institute with an account of my recent researches and discoveries, the appearance of a book coming from England, and containing a description of a number of experiments and discoveries in many respects not dissimilar to my own, could not fail to create a certain feeling of surprise among the philosophers of Paris, as I could plainly enough perceive. . . .

"I am far from intending to assert that Mr. Leslie had any knowledge of those experiments of mine which bore a resemblance to those which he announced publicly in print. It is, however, equally certain that I did not know, and could not have known, the least thing about his. . . .

"As regards the *priority of the public announcement* of our discoveries, this point can be easily made clear by the statement of certain facts which do not admit of doubt.

"It is true that I cannot determine with any great accuracy the time when Mr. Leslie's book first saw the light; it cannot, however, possibly have been published before the middle of May of this year, for the dedication is dated at Largo, in Fifeshire (Scotland), the 20th of May, 1804. This would be, consequently, nearly a year after the time when the most remarkable results of my investigations were known in London."

Rumford's published proof that he could not have designed his experiments after Leslie's did not, however, still the clamor of accusations. In fact it reached proportions that even ladies of fashionable society entered the fray. We find the widow of the celebrated chemist Lavoisier writing to the Swiss physicist Pictet,† in June 1805:

"I presume, Sir, that you have received the *Monthly Review* for the months of March and April, and that you have noted the manner in which Count Rumford is treated by this publication. . . . Your friendship for Count

† Unpublished letter, transcript in the collection of the American Academy of Arts and Sciences, Boston.

Rumford will certainly revolt at reading such an article; and don't you think it proper for you to come to his defense. . . . I am expressing to you, Sir, my idea and my desire to see an atrocious calamity repulsed . . . injustice pains true men, and you were certainly not in need of my warning to make you come to the Count's aid."†

Active interest in the controversy gradually died with the general acceptance in England of Leslie as the discoverer of the differential thermometer, while on the Continent Rumford was given the credit. However, the whole debate was reopened when Sir Humphrey Davy published his *Elements of Chemical Philosophy* in 1812. The issues of the new controversy can best be described by Leslie himself.‡

"In glancing over Davy's *Elements of Chemical Philosophy*, I was surprised to find it alleged that Van Helmont had given a sketch of a 'curious instrument, very similar to the differential thermometer;' and a few pages further on, to meet with this bottom note—'Plate [12] represents Mr. Leslie's differential thermometer. Plate [13] is copied from Van Helmont. This instrument appears to have been the first in which the expansive power of heated air was exhibited by its action upon cold air.' On turning to this plate, I observe a very clumsy and distorted representation of my differential thermometer, placed by the side of another figure, bearing a general resemblance to it, and purporting to be a copy of Van Helmont's sketch. I soon perceived, however, that Van Helmont's description and figure [see Plate 14] were essentially different from the representation which Sir Humphrey has taken the trouble to give. In fact, the 'curious instrument' described by the alchemist is no other than the common air-thermometer of Sanctorio or Drebbel§, invented more than forty years before his death; only, for the sake of easier carriage, shaped like a siphon, the lower end being bent upwards, and terminating in a spherical cup, with a *small orifice*—one of the forms which it had from its earliest introduction. A learned person of the name of Heer imagining, it seems, as Sir Humphrey has since done, that the instrument was absolutely closed, had proceeded to admire the perpetual motion of the contained liquid. . . . To this Van Helmont replied, that the action of the machine no more produced perpetual motion than the changing of a weathercock . . . he calls his rival, in the uncourtly language of those times, 'an idiot', and charges him with 'stupidity', for not perceiving that the instrument had an *aperture*, only slightly shut with a stopper, and not hermetically sealed, as it is most incorrectly figured in the *Elements of Chemical Philosophy*.

"I will not suppose that Sir Humphrey intentionally misrepresented the meaning of Van Helmont; but then it follows, that either he had not read the passage to which he refers, or must have satisfied himself with a very

† Since Mme Lavoisier married Rumford four months later, she may have had a more than casual interest in the "atrocious calamity" which was befalling him.

‡ J. Leslie, *Phil. Mag.* **4**, 236 (1812).

§ Boerhaave, *Elementa Cheminae*, Paris (1733) p. 83.

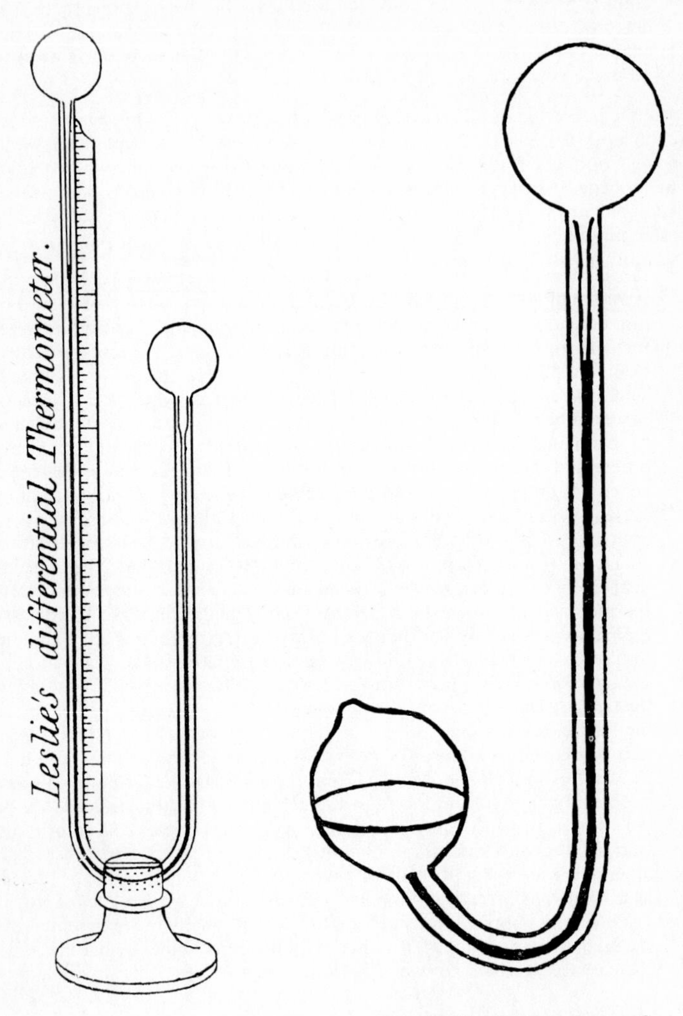

PLATE 12. Davy's copy of Leslie's differential thermometer.

PLATE 13. Davy's copy of Van Helmont's thermometer.

superficial and careless inspection. This precipitancy is the more to be lamented, as it may possibly beget suspicion of want or accuracy or fairness in other matters of higher consequence."

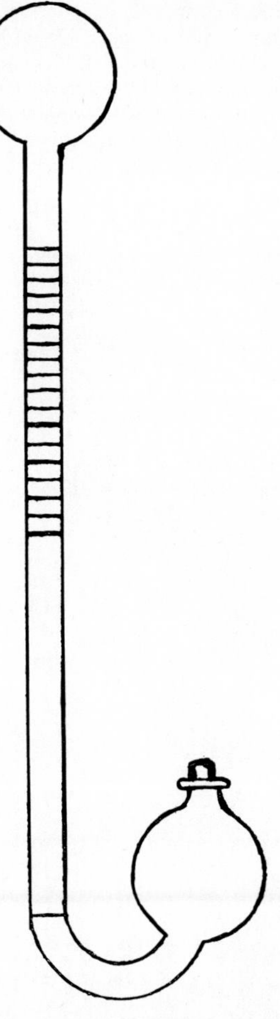

PLATE 14. Leslie's copy of Van Helmont's thermometer.

Following this original complaint of Leslie against Davy, the *Philosophical Magazine* for 1812 and 1813 contains several articles on both sides of this argument. Both Sir John and Sir Humphrey were accused of falsifying illustrations, suppressing facts, and wilful misrepresentations!*

Controversies of this sort are interesting not so much for the physics involved as for the light they shed on the physicists. Sir John Leslie was involved in many such battles, and the sarcastic insinuation for which he was noted is well illustrated in the passage quoted above. Count Rumford displayed further his egotistical personality during the first dispute when he wrote:† "The *priority* in question, considered in and by itself, is of such slight importance that I should not have mentioned it at all, were it not that the facts which go to establish it tend at the same time to strengthen a far more important assertion, namely, that I am actually the *discoverer* of what I announced as discoveries."

* See, for example, "A. B." *Phil. Mag.* **40,** 329 (1812) and "C. D." *ibid.* **41** 31 (1813).

† Unpublished letter, transcript in the collection of the American Academy of Arts and Sciences, Boston.

CHAPTER 11

COUNT RUMFORD'S CONCEPT OF HEAT†

CLIMAXING his famous paper,‡ "An Inquiry Concerning the Source of the Heat which is Excited by Friction", Count Rumford wrote: "It appears to me to be extremely difficult, if not quite impossible, to form any distinct idea of anything capable of being excited and communicated in the manner the Heat was excited and communicated in these experiments, except it be MOTION."

It is of interest to inquire what Rumford meant by "motion" in this often-quoted passage, and to find that intuitively he accepted an interesting mixture of a remarkably modern theory of heat in solids with the concept that temperature was associated with the frequency of an oscillating source.

When Benjamin Thompson was seventeen years of age, and furthering his education as best he could by reading everything available, he came across Boerhaave's *Treatise on Fire.*§ Boerhaave, whose books on natural philosophy were standard texts in the eighteenth century, gave great weight to the theory that heat, like sound, was a product of the vibration of the heated body. Thompson accepted this point of view very early in his life. Almost never did he give credit to previous scientific work, but one of the very few references he did cite* in his many volumes of writings

† Reproduced, with permission, from the *American Journal of Physics* **20,** 331–4 (1952).

‡ Count Rumford, *Trans. Roy. Soc.* (*London*) **88,** 80 (1798).

§ Count Rumford, *Works of Count Rumford* (American Academy of Arts and Sciences, Boston, 1873), vol. II, p. 188.

* B. Thompson, *Trans. Roy. Soc.* (*London*) **71,** 229 (1781).

was to Boerhaave's *Chemistry* in which this author likens a heated body to a struck bell, an analogy which the Count uses constantly.†

> Suppose now that heat be nothing more than the motions of the constituent particles of bodies among themselves (an hypothesis of ancient date, and which always appeared to me to be very probable), if for the bell we substitute a hot body, the cooling of it will be attended by a series of actions and reactions exactly similar to those just described.
>
> The rapid undulations occasioned in the surrounding ethereal fluid, by the swift vibrations of the hot body, will act as calorific rays on the neighbouring colder solid bodies, and the slower undulations, occasioned by the vibrations of those colder bodies, will act as frigorific rays on the hot body; and these reciprocal actions will continue, but with decreasing intensity, till the hot body and those colder bodies which surround it shall, in consequence of these actions, have acquired the same temperature, or until their vibrations have become isochronous.

The experiments which led Thompson to the vibratory theory of heat were those which he performed in England in 1778 on the heating of guns during firing.‡ He found that guns fired without bullets always got hotter than those which fired shot. In attempting to explain these observations he reasoned that the expanding gases resulting from the explosions passed through the gun with a higher velocity when they were free than when they were impelling a ball before them. He went on to conclude that the higher-velocity explosion produced a higher-frequency oscillation of the material of the gun, and hence a higher temperature.

The ordinary phenomena associated with heat were easily explained by this theory. Thermal expansion and contraction resulted from an increased or decreased amplitude of oscillation of the constituent particles which made up the body. The oscillatory theory postulated a maximum amplitude of oscillation of the particles that made up a body which, if exceeded, caused a change in phase. The latent heat given up when a liquid froze was taken as proof that the vibratory motion resided also in the liquid and became available for measurement when the liquid solidified.

† Count Rumford, *Trans. Roy. Soc.* (*London*) **94,** 77 (1804).

‡ B. Thompson, *Trans. Soc.* (*London*) **71,** 229 (1781).

Incidentally, Rumford rejected the concept of an absolute zero of temperature with the following words:†

> *Hot* and *cold*, like *fast* and *slow*, are mere relative terms; and, as there is no relation or proportion between motion and a state of rest, so there can be no relation between any degree of heat and absolute cold, or a total privation of heat; hence it is evident that all attempts to determine the place of *absolute cold*, on the scale of a thermometer, must be nugatory. It seems probable that *motion* is an essential quality of matter and that rest is nowhere to be found in the universe.

One of the attractive features of the vibratory theory was that the same mechanism applied in conduction as in radiation of heat. He wrote:‡ "A theory which should have the advantage of explaining the communication of heat by a *single* method, at once simple and easy to understand, would be preferable, it seems to me, to one which, in order to explain various phenomena, would be obliged to admit *two different* modes of the communication of heat." Rumford recognized no difference between acoustic and optic waves and was led to make the shrewd observation that§

> there are so many striking analogies between the rays of light and those invisible rays which all bodies at all temperatures appear to emit, that we can hardly doubt of their motions being regulated by the same laws. Perhaps there may be no other difference between them than exists between those vibrations in the air which are audible and those which make no sensible impression on our organs of hearing.

The Count's concept of heat is well illustrated in his introduction to his cannon boring experiment:*

> I argued that if the existence of caloric was a fact, it must be absolutely impossible for a body or for several individual bodies, which together made one whole, to communicate this substance continuously to various other bodies by which they were surrounded, without this substance gradually being entirely exhausted. A sponge filled with water and hung

† Count Rumford, *Trans. Roy. Soc.* (*London*) **94,** 77 (1804).

‡ Count Rumford, *Moniteur Universal*, 9 Messidor, An 12 (June 26, 1804).

§ Count Rumford, *Works of Count Rumford* (American Academy of Arts and Sciences, Boston, 1873), vol. II, p. 188.

* Count Rumford, *Works of Count Rumford* (American Academy of Arts and Sciences, Boston, 1873), vol. II, p. 188.

by a thread in the middle of a room filled with dry air, communicates its moisture to the air, it is true, but soon the water evaporates and the sponge can no longer give out moisture. On the contrary, a bell sounds without interruption when it is struck, and gives out its sound as often as we please without the slightest perceptible loss. Moisture is a substance; sound is not. It is well known that two hard bodies, if rubbed together, produce much heat. Can they continue to produce it without finally becoming exhausted? Let the result of experiment decide this question.

Rumford's conviction that each temperature corresponded to a particular frequency of oscillation of a body led him to champion actively the concept of "calorific" and "frigorific" rays first postulated by Plutarch in his *De Primo Frigido*. Ever since Jean Baptista Porta in 1590 conducted experiments showing the focusing of radiation from cold bodies by concave mirrors, those who repeated these experiments found it difficult not to admit the existence of frigorific rays. During the 1780's Rumford's intimate friend and fellow philosopher, Professor Pictet of Geneva, carried out a long series of experiments on the focusing of cold which so intrigued the Count that he also conducted many tests to show that frigorific radiation existed.

In the autumn of 1800, while visiting Scotland, Rumford, with three professors from the University of Edinburgh (Hope, Playfair, and Stewart) conducted a typical experiment of this kind:†

Two metallic mirrors fifteen inches in diameter, with a focal distance of fifteen inches, were placed opposite each other, sixteen feet apart. When a cold body (for example, a glass bulb filled with water and pounded ice) as was the case on this occasion, was placed in the focus of one of the mirrors, and a very sensitive air-thermometer was placed in the focus of the other mirror, the latter thermometer began immediately to fall. If, instead of being placed directly in the focus, the thermometer was removed a short distance from it to one side, the cooling power which in the former case the cold body had exerted upon it was no longer perceptible.

As Rumford wrote‡ to Pictet in Geneva on October 18, 1800: "We repeated your interesting experiment on the reflection of cold,

† Count Rumford, *Works of Count Rumford* (American Academy of Arts and Sciences, Boston, 1873), vol. II, p. 188.

‡ Manuscript Collection of the American Academy of Arts and Sciences, Boston, Massachusetts.

PLATE 15. Royal Institution mirrors for focusing heat and cold.

two days ago at Dr. Hope's house, and with complete success. . . . The slow vibrations of ice in the bottle cause the thermometer to sing a lower note."

Count Rumford equipped his Royal Institution with four large mirrors for demonstrating the focusing of "calorific" and "frigorific" rays. The use of these is illustrated in the figure taken from John Tyndall's *Heat Considered as a Mode of Motion.*† Tyndall was Professor of Natural Philosophy in the Royal Institution and his demonstration lectures in physics have survived for almost a hundred years as a model for many similar survey courses. In the figure he is shown exploding a hydrogen–oxygen balloon by means of Rumford's mirrors, but in the next experiment he says:

> I again lower the mirror, and, . . . suspend a [flask] containing a freezing mixture. I raise the mirror and, as in the former case, bring the [thermo] pile into the focus of the lower one. Turned directly toward the upper flask there is no action; turned downwards, the needle moves: observe the direction of the motion—the red end curves toward me [indicating cold]. Does it not appear as if this body in the upper focus were now emitting rays of cold which are converged by the lower mirror exactly as the rays of heat in our former experiment. The facts are exactly complementary, and it would seem that we have precisely the same right to infer from the experiments, the existence and convergence of these cold rays, as we have to infer the existence and convergence of the heat rays.

Count Rumford's belief that cold was a separate entity from heat was a very real one. In discussing the radiation of cold he wrote:‡ "According to this hypothesis, *cold* can with no more propriety be considered as the absence *of heat* than a low or grave sound can be considered as the absence of a higher or more acute note; and the admission of rays which generate cold involve no absurdity and create no confusion of ideas."

He saw in the workings of nature confirmation of his ideas. Were not people black in hot climates and white in cold ones? The black skin of the Negro allowed him to radiate efficiently and thereby keep cool, while white skin was an efficient reflector of

† John Tyndall, *Heat Considered as a Mode of Motion* (D. Appleton-Century Company, Inc., New York, 1871).

‡ Count Rumford, *Trans. Roy. Soc.* (*London*) **94,** 77 (1804).

frigorific radiation and hence defended the white man from the cold. Rumford was so sure of his conclusions that he carried his convictions to the logical conclusion of always wearing white clothing in cold weather, much to the derisive amusement of the Parisian society in which he moved in later years.

Rumford had no concept of heat as a random motion. He felt that heat was primarily set up by the harmonic vibrations of the "fibers of the metal,"† was transmitted through solids and radiated from them in the same manner as acoustic waves. He did not feel that these same waves could be set up in fluids. In fact, he carried out a long series of experiments‡ showing that gases and liquids (including mercury!) were perfect nonconductors and that their only mode of communicating heat was by convection.§ He felt that what heat was transmitted through fluids at rest was due only to the conduction of thermal vibrations in the all-pervading ether, and he was strengthened in this belief by his showing that heat passed almost as easily through a Torricellian vacuum as through air.* By 1800 he was completely convinced that heat was a vibratory motion, analogous in every way to acoustical oscillations.

† B. Thompson, *Trans. Roy. Soc.* (*London*) **71,** 229 (1781).
‡ Count Rumford, *Essay VII*, *Part II*, Cadel and Davies, London, 1798.
§ S. C. Brown, *Am. J. Phys.* **15,** 273 (1947).
* Count Rumford, *Trans. Roy. Soc.* (*London*) **76,** 273 (1786).

INDEX